ARMY, MARINE CORPS, NAVY, AIR FORCE

SURVIVAL, EVASION, AND RECOVERY

MULTISERVICE PROCEDURES FOR SURVIVAL, EVASION, AND RECOVERY

FM 21-76-1
MCRP 3-02H
NWP 3-50.3
AFTTP(I) 3-2.26

JUNE 1999

AIR LAND SEA APPLICATION CENTER

DISTRIBUTION RESTRICTION:
Approved for public release; distribution is unlimited.

MULTISERVICE TACTICS, TECHNIQUES, AND PROCEDURES

QUICK REFERENCE CHECKLIST

Decide to Survive!

S	-	**S**ize up the situation, surroundings, physical condition, equipment.
U	-	**U**se all your senses
R	-	**R**emember where you are.
V	-	**V**anquish fear and panic.
I	-	**I**mprovise and improve.
V	-	**V**alue living.
A	-	**A**ct like the natives.
L	-	**L**ive by your wits.

1. Immediate Actions

a. Assess immediate situation. ***THINK BEFORE YOU ACT!***

b. Take action to protect yourself from nuclear, biological, or chemical hazards (Chapter IX).

c. Seek a concealed site.

d. Assess medical condition; treat as necessary (Chapter V).

e. Sanitize uniform of potentially compromising information.

f. Sanitize area; hide equipment you are leaving.

g. Apply personal camouflage.

h. Move away from concealed site, zigzag pattern recommended.

i. Use terrain to advantage, communication, and concealment.

j. Find a hole-up site.

2. Hole-Up-Site (Chapter I)

a. Reassess situation; treat injuries, then inventory equipment.

b. Review plan of action; establish priorities (Chapter VI).

c. Determine current location.

d. Improve camouflage.

e. Focus thoughts on task(s) at hand.

f. Execute plan of action. Stay flexible!

Recommend inclusion of this manual in the aviator's survival vest.

3. Concealment (Chapter I)

a. Select a place of concealment providing—

(1) Adequate concealment, ground and air.

(2) Safe distance from enemy positions and lines of communications (LOC).

(3) Listening and observation points.

(4) Multiple avenues of escape.

(5) Protection from the environment.

(6) Possible communications/signaling opportunities.

b. Stay alert, maintain security.

c. Drink water.

4. Movement (Chapters I and II)

a. Travel slowly and deliberately.

b. ***DO NOT*** leave evidence of travel; use noise and light discipline.

c. Stay away from LOC.

d. Stop, look, listen, and smell; take appropriate action(s).

e. Move from one concealed area to another.

f. Use evasion movement techniques (Chapter I).

5. Communications and Signaling (Chapter III)

a. Communicate as directed in applicable plans/orders, particularly when considering transmitting ***in the blind***.

b. Be prepared to use communications and signaling devices on short notice.

c. Use of communications and signaling devices may compromise position.

6. Recovery (Chapter IV)

a. Select site(s) IAW criteria in theater recovery plans.

b. Ensure site is free of hazards; secure personal gear.

c. Select best area for communications and signaling devices.

d. Observe site for proximity to enemy activity and LOC.

e. Follow recovery force instructions.

FM 21-76-1 U.S. Army Training and Doctrine Command
Fort Monroe, Virginia

MCRP 3-02H Marine Corps Combat Development Command
Quantico, Virginia

NWP 3-50.3 Navy Warfare Development Command
Newport, Rhode Island

AFTTP(I) 3-2.26 Headquarters Air Force Doctrine Center
Maxwell Air Force Base, Alabama

29 JUNE 1999

Survival, Evasion, and Recovery

Multiservice Procedures for Survival, Evasion, and Recovery

Note: This *UNCLASSIFIED* publication is designed to provide Service members quick-reference survival, evasion, and recovery information. See Appendix B for the scope, purpose, application, implementation plan, and user information.

TABLE OF CONTENTS

THE CODE OF CONDUCT

ARTICLE I

I am an American, fighting in the forces which guard my country and our way of life. I am prepared to give my life in their defense.

ARTICLE II

I will never surrender of my own free will. If in command, I will never surrender the members of my command while they still have the means to resist.

ARTICLE III

If I am captured, I will continue to resist by all means available. I will make every effort to escape and aid others to escape. I will accept neither parole nor special favors from the enemy.

ARTICLE IV

If I become a prisoner of war, I will keep faith with my fellow prisoners. I will give no information or take part in any action which might be harmful to my comrades. If I am senior, I will take command. If not, I will obey the lawful orders of those appointed over me and will back them up in every way.

ARTICLE V

When questioned, should I become a prisoner of war, I am required to give name, rank, service number and date of birth. I will evade answering further questions to the utmost of my ability. I will make no oral or written statements disloyal to my country and its allies or harmful to their cause.

ARTICLE VI

I will never forget that I am an American, fighting for freedom, responsible for my actions, and dedicated to the principles which made my country free. I will trust in my God and in the United States of America.

Chapter I
EVASION

1. Planning

a. Review the quick reference checklist on the inside cover.

b. Guidelines for successful evasion include—
   (1) Keeping a positive attitude.
   (2) Using established procedures.
   (3) Following your evasion plan of action.
   (4) Being patient.
   (5) Drinking water (***DO NOT*** eat food without water).
   (6) Conserving strength for critical periods.
   (7) Resting and sleeping as much as possible.
   (8) Staying out of sight.

c. The following odors stand out and may give an evader away:
   (1) Scented soaps and shampoos.
   (2) Shaving cream, after-shave lotion, or other cosmetics.
   (3) Insect repellent (camouflage stick is least scented).
   (4) Gum and candy (smell is strong or sweet).
   (5) Tobacco (odor is unmistakable).

d. Where to go (initiate evasion plan of action):
   (1) Near a suitable area for recovery.
   (2) Selected area for evasion.
   (3) Neutral or friendly country or area.
   (4) Designated area for recovery.

2. Camouflage

a. Basic principles:
   (1) Disturb the area as little as possible.
   (2) Avoid activity that reveals movement to the enemy.
   (3) Apply personal camouflage.

b. Camouflage patterns **(Figure I-1)**:
   (1) Blotch pattern.
      (a) Temperate deciduous (leaf shedding) areas.
      (b) Desert areas (barren).
      (c) Snow (barren).
   (2) Slash pattern.
      (a) Coniferous areas (broad slashes).

(b) Jungle areas (broad slashes).
(c) Grass (narrow slashes).
(3) Combination. May use blotched and slash together.

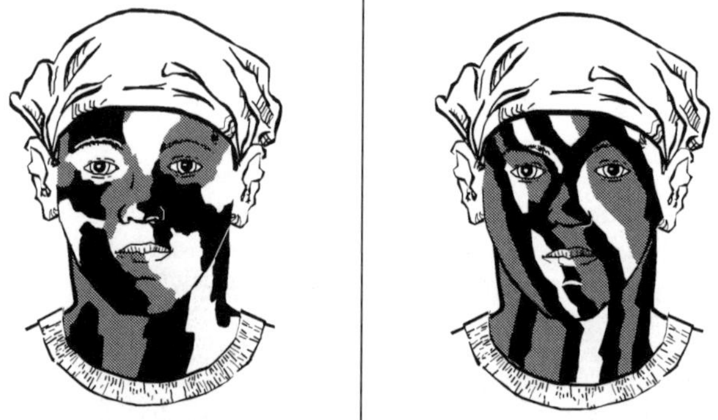

Figure I-1. Camouflage Patterns

c. Personal camouflage application follows:

(1) Face. Use dark colors on high spots and light colors on any remaining exposed areas. Use a hat, netting, or mask if available.

(2) Ears. The insides and the backs should have **2** colors to break up outlines.

(3) Head, neck, hands, and the under chin. Use scarf, collar, vegetation, netting, or coloration methods.

(4) Light colored hair. Give special attention to conceal with a scarf or mosquito head net.

d. Position and movement camouflage follows:
(1) Avoid unnecessary movement.
(2) Take advantage of natural concealment:
(a) Cut foliage fades and wilts, change regularly.
(b) Change camouflage depending on the surroundings.
(c) ***DO NOT*** select vegetation from same source.
(d) Use stains from grasses, berries, dirt, and charcoal.
(3) ***DO NOT*** over camouflage.
(4) Remember when using shadows, they shift with the sun.

(5) Never expose shiny objects (like a watch, glasses, or pens).
(6) Ensure watch alarms and hourly chimes are turned off.
(7) Remove unit patches, name tags, rank insignia, etc.
(8) Break up the outline of the body, ***"V"*** of crotch/armpits.
(9) Conduct observation from a prone and concealed position.

3. Shelters

a. Use camouflage and concealment.
b. Locate carefully—easy to remember acronym: ***BLISS***.

B - **B**lend **L** - **L**ow silhouette **I** - **I**rregular shape **S** - **S**mall **S** - **S**ecluded location

(1) Choose an area—
(a) Least likely to be searched (drainages, rough terrain, etc.) and blends with the environment.
(b) With escape routes (***DO NOT*** corner yourself).
(c) With observable approaches.
(2) Locate entrances and exits in brush and along ridges, ditches, and rocks to keep from forming paths to site.
(3) Be wary of flash floods in ravines and canyons.
(4) Conceal with minimal to no preparation.
(5) Take the direction finding threat into account before transmitting from shelter.
(6) Ensure overhead concealment.

4. Movement

a. A moving object is easy to spot. If travel is necessary—
(1) Mask with natural cover **(Figure I-2)**.
(2) Use the military crest.
(3) Restrict to periods of low light, bad weather, wind, or reduced enemy activity.

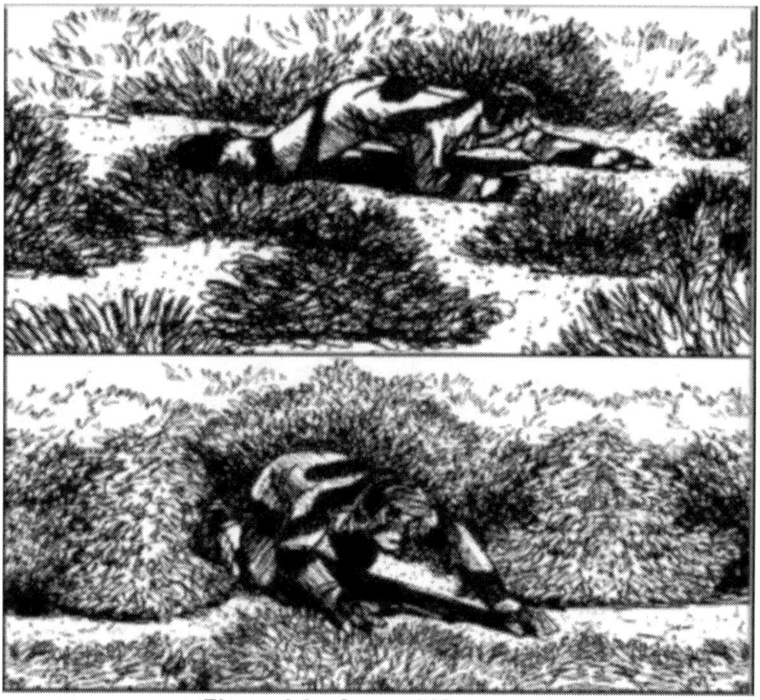

Figure I-2. Ground Movement

(4) Avoid silhouetting **(Figure I-3)**.

(5) At irregular intervals—

(a) ***STOP*** at a point of concealment.

(b) ***LOOK*** for signs of human or animal activity (smoke, tracks, roads, troops, vehicles, aircraft, wire, buildings, etc.). Watch for trip wires or booby traps and avoid leaving evidence of travel. Peripheral vision is more effective for recognizing movement at night and twilight.

(c) ***LISTEN*** for vehicles, troops, aircraft, weapons, animals, etc.

(d) ***SMELL*** for vehicles, troops, animals, fires, etc.

Figure I-3. Avoid Silhouetting

(6) Employ noise discipline; check clothing and equipment for items that could make noise during movement and secure them.

b. Break up the human shape or recognizable lines.

c. Route selection requires detailed planning and special techniques (irregular route/zigzag) to camouflage evidence of travel.

d. Some techniques for concealing evidence of travel follows:

(1) Avoid disturbing the vegetation above knee level.

(2) ***DO NOT*** break branches, leaves, or grass.

(3) Use a walking stick to part vegetation and push it back to its original position.

(4) ***DO NOT*** grab small trees or brush. (This may scuff the bark or create movement that is easily spotted. In snow country, this creates a path of snowless vegetation revealing your route.)

(5) Pick firm footing (carefully place the foot lightly but squarely on the surface to avoid slipping). ***TRY NOT TO—***

(a) Overturn ground cover, rocks, and sticks.

(b) Scuff bark on logs and sticks.

(c) Make noise by breaking sticks. (Cloth wrapped around feet helps muffle this.)

(d) Mangle grass and bushes that normally spring back.

(6) Mask unavoidable tracks in soft footing by—

(a) Placing tracks in the shadows of vegetation, downed logs, and snowdrifts.

(b) Moving before and during precipitation allows tracks to fill in.

(c) Traveling during windy periods.

(d) Taking advantage of solid surfaces (logs, rocks, etc.) leaving less evidence of travel.

(e) Patting out tracks lightly to speed their breakdown or make them look old.

(7) Secure trash or loose equipment—hide or bury discarded items. (Trash or lost equipment identifies who lost it.)

(8) Concentrate on defeating the handler if pursued by dogs.

e. Penetrate obstacles as follows:

(1) Enter deep ditches feet first to avoid injury.

(2) Go around chain-link and wire fences. Go under fence if unavoidable, crossing at damaged areas. ***DO NOT*** touch fence; look for electrical insulators or security devices.

(3) Penetrate rail fences, passing under or between lower rails. If impractical, go over the top, presenting as low a silhouette as possible **(Figure I-4)**.

(4) Cross roads after observation from concealment to determine enemy activity. Cross at points offering concealment such as bushes, shadows, bend in road, etc. Cross in a manner leaving your footprints parallel (cross step sideways) to the road. **(Figure I-5)**

(5) Use same method of observation for railroad tracks that was used for roads. Next, align body parallel to tracks with face down, cross tracks using a semi-pushup motion. Repeat for the second track. **(Figure I-6)**.

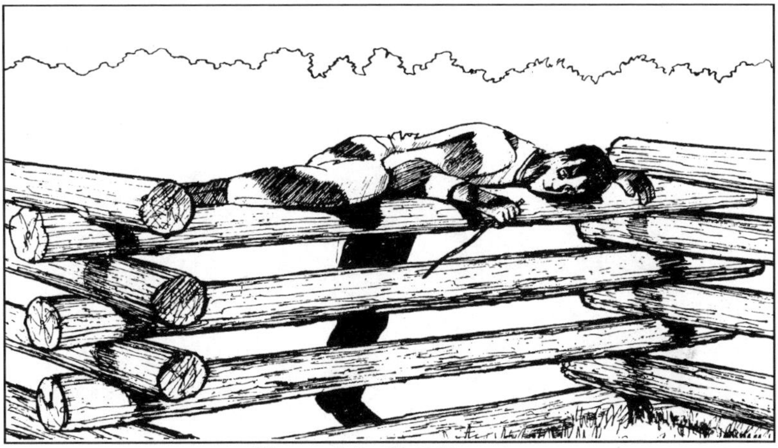

Figure I-4. Rail Fences

Figure I-5. Road Crossing

Figure I-6. Railroad Tracks

WARNING: If 3 rails exist, 1 may be electrified.

Chapter II

NAVIGATION

Assess the threat and apply appropriate evasion principles.

1. Stay or Move Considerations

a. Stay with the vehicle/aircraft in a non-combat environment.

b. Leave only when—

(1) Dictated by the threat.

(2) Are certain of your location, have a known destination, and have the ability to get there.

(3) Can reach water, food, shelter, and/or help.

(4) Convinced rescue is not coming.

c. Consider the following if you decide to travel:

(1) Follow the briefed evasion plan.

(2) Determine which direction to travel and why.

(3) Decide what equipment to take, cache, or destroy.

d. Leave information at your starting point (in a non-combat environment) that includes—

(1) Destination.

(2) Route of travel.

(3) Personal condition.

(4) Supplies available.

e. Consider the following for maps (in a combat environment):

(1) ***DO NOT*** write on the map.

(2) ***DO NOT*** soil the map by touching the destination.

(3) ***DO NOT*** fold in a manner providing travel information.

Note: **These actions may compromise information if captured.**

2. Navigation and Position Determination

a. Determine your general location by—

(1) Developing a working knowledge of the operational area.

(a) Geographic checkpoints.

(b) Man-made checkpoints.

(c) Previous knowledge of operational area.

(2) Using the ***Rate x Time = Distance*** formula.

(3) Using information provided in the map legend.

(4) Using prominent landmarks.

(5) Visualizing map to determine position.

b. Determine cardinal directions (north, south, east, and west) by—

(1) Using compass.

CAUTION: The following methods are ***NOT*** highly accurate and give only general cardinal direction.

(2) Using stick and shadow method to determine a true north-south line **(Figure II-1)**.

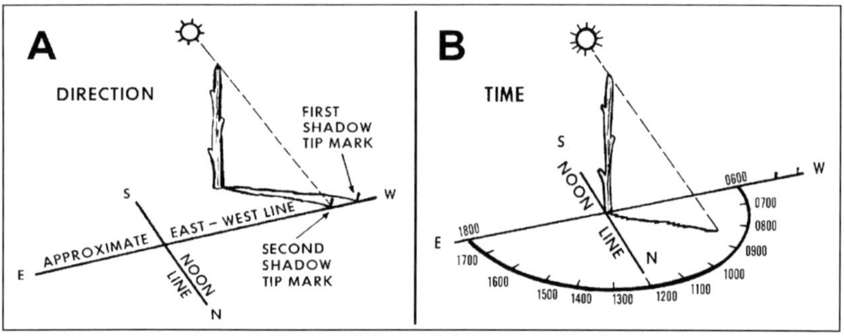

Figure II-1. Stick and Shadow Method

(3) Remembering the sunrise/moonrise is in the east and sunset/moonset is in the west.

(4) Using a wristwatch to determine general cardinal direction **(Figure II-2)**.

(a) Digital watches. Visualize a clock face on the watch.

(b) Northern Hemisphere. Point hour hand at the sun. South is halfway between the hour hand and 12 o'clock position.

(c) Southern Hemisphere. Point the 12 o'clock position on your watch at the sun. North is halfway between the 12 o'clock position and the hour hand.

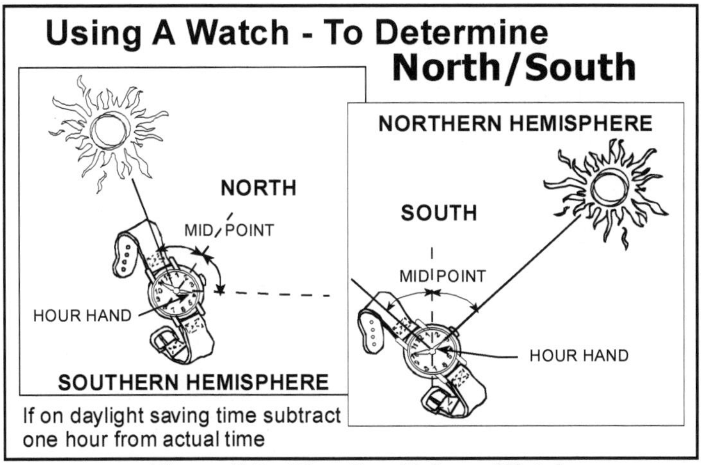

Figure II-2. Direction Using a Watch

(5) Using a pocket navigator **(Figure II-3)**—

(a) Gather the following necessary materials:

•Flat writing material (such as an MRE box).

•1-2 inch shadow tip device (a twig, nail, or match).

•Pen or pencil.

(b) Start construction at sunup; end construction at sundown. Do the following:

•Attach shadow tip device in center of paper.

•Secure navigator on flat surface (***DO NOT*** move during set up period).

•Mark tip of shadow every 30 minutes annotating the time.

•Connect marks to form an arc.

•Indicate north with a drawn arrow.

Note: The shortest line between base of shadow tip device and curved line is a north-south line.

(c) Do the following during travel:

•Hold navigator so the shadow aligns with mark of present time (drawn arrow now points to true north).

(d) Remember the navigator is current for approximately 1 week.

CAUTION: The Pocket Navigator is ***NOT*** recommended if evading.

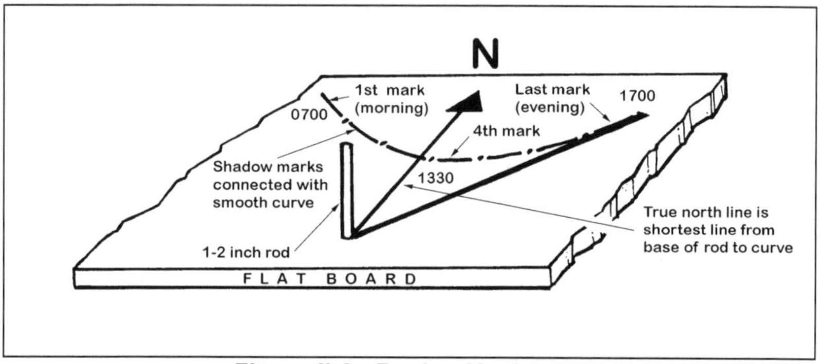

Figure II-3. Pocket Navigator

(6) Using the stars **(Figure II-4)** the—

(a) North Star is used to locate true north-south line.

(b) Southern Cross is used to locate true south-north line.

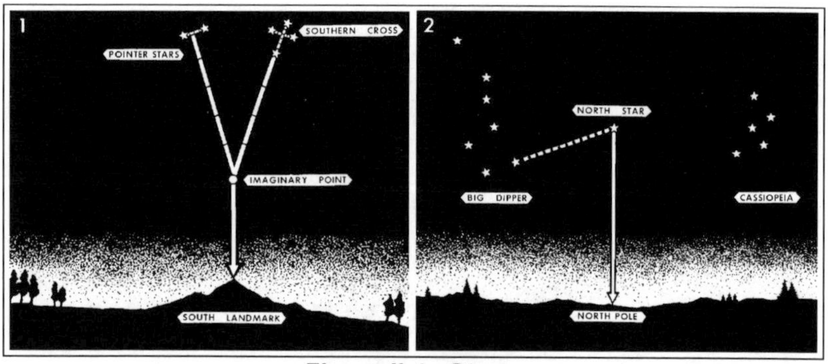

Figure II-4. Stars

c. Orient the map by—

(1) Using a true north-south line **(Figure II-5)**—

(a) Unfold map and place on a firm, flat, level nonmetallic surface.

(b) Align the compass on a true north-south line.

(c) Rotate map and compass until stationary index line aligns with the magnetic variation indicated in marginal information.

•Easterly (subtract variation from 360 degrees).

•Westerly (add variation to 360 degrees).

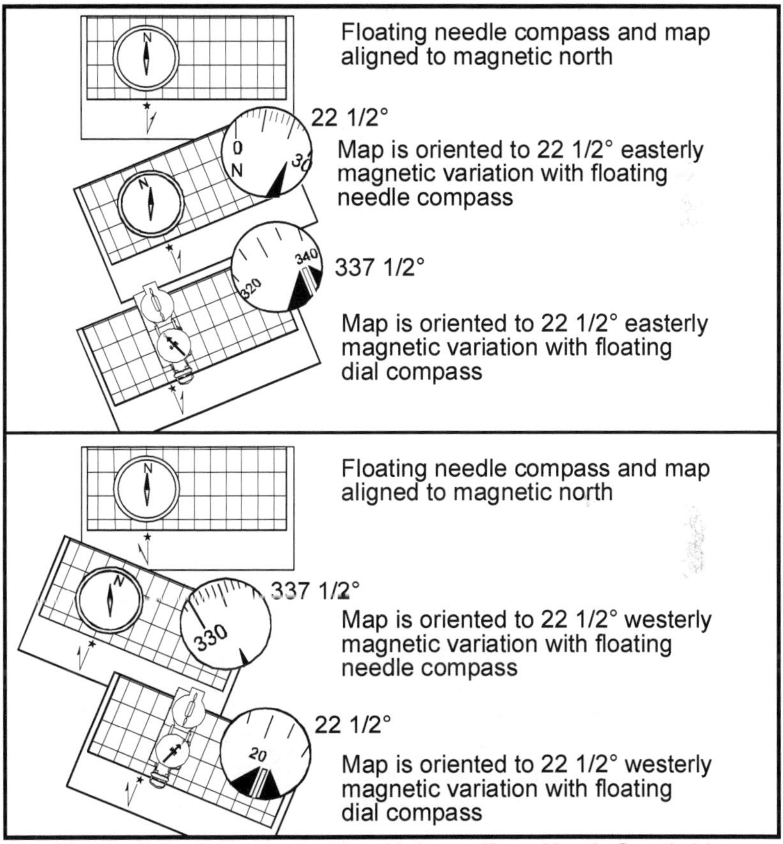

Figure II-5. Orienting a Map Using a True North-South Line

(2) Using a compass rose **(Figure II-6)**—

(a) Place edge of the lensatic compass on magnetic north line of the compass rose closest to your location.

(b) Rotate map and compass until compass reads 360 degrees.

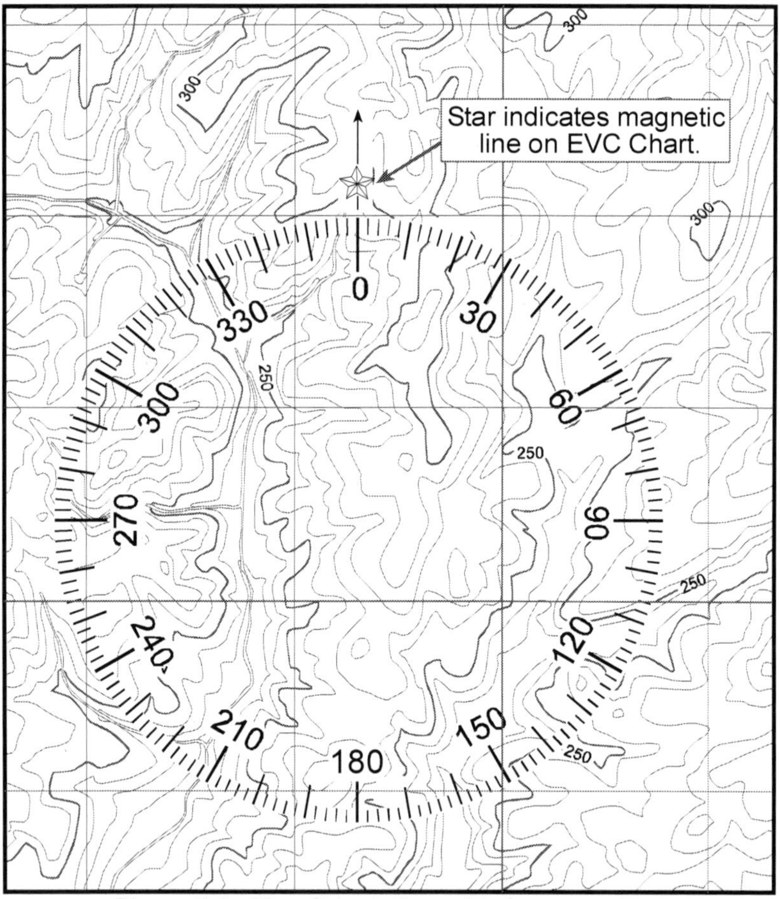

Figure II-6. Map Orientation with Compass Rose

(3) If there is **NO** compass, orient map using cardinal direction obtained by the stick and shadow method or the celestial aids (stars) method.

d. Determine specific location.

(1) Global Positioning System (GPS).

(a) ***DO NOT*** use GPS for primary navigation.

(b) Use GPS to confirm your position ***ONLY***.

(c) Select area providing maximum satellite reception.

(d) Conserve GPS battery life.

(2) Triangulation (resection) with a compass **(Figure II-7)**.

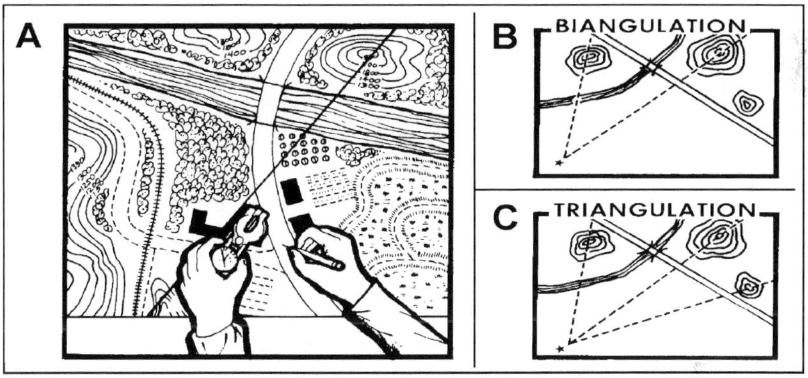

Figure II-7. Triangulation

(a) Try to use **3** or more azimuths.

(b) Positively identify a major land feature and determine a line of position (LOP).

(c) Check map orientation each time compass is used.

(d) Plot the LOP using a thin stick or blade of grass (combat) or pencil line (non-combat).

(e) Repeat steps **(b)** through **(d)** for other LOPs.

e. Use the compass for night navigation by—

(1) Setting up compass for night navigation **(Figure II-8)**.

(2) Aligning north-seeking arrow with luminous line and follow front of compass.

(3) Using point-to-point navigation.

f. Route selection techniques follow:

Setting the Compass for Night Travel

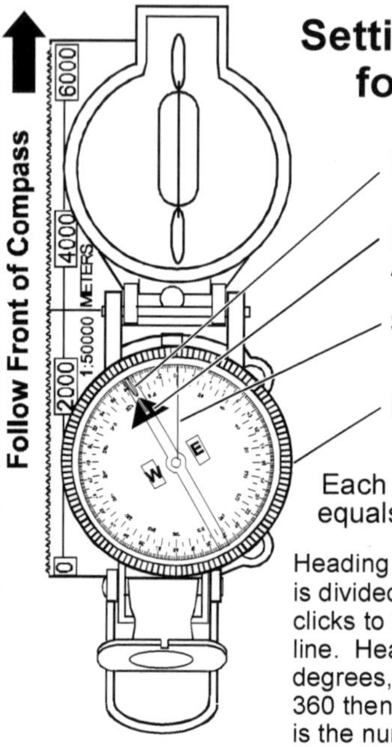

Each click of the Bezel Ring equals 3 degrees.

Heading between 0 and 180 degrees is divided by 3. Sum is number of clicks to the left of stationary index line. Heading between 180 and 360 degrees, subtract heading from 360 then divide sum by 3. New sum is the number of clicks to the right from stationary index line.

EXAMPLES

Heading of 027 degrees = 9 clicks left.
Heading of 300 degrees = 20 clicks right.

Figure II-8. Compass Night Navigation Setup

(1) Circumnavigation.

(a) Find a prominent landmark on the opposite side of the obstacle.

(b) Contour around obstacle to landmark.

(c) Resume your route of travel.

(2) Dogleg and 90 degree offset **(Figure II-9)**.

(3) Straight-line heading as follows:

(a) Maintain heading until reaching destination.

(b) Measure distance by counting the number of paces in a given course and convert to map units.

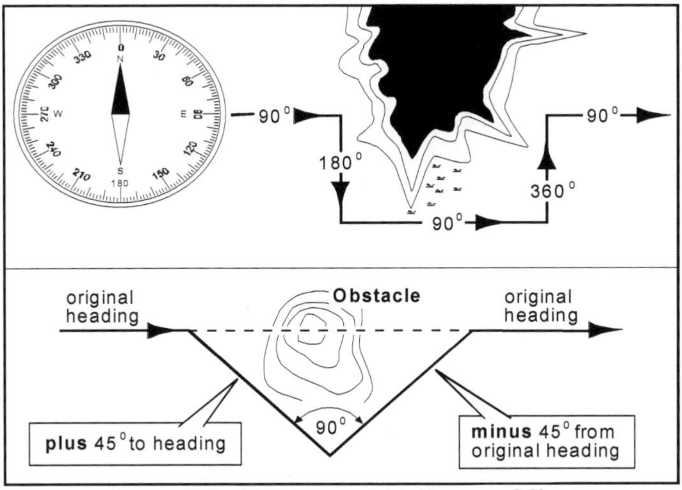

Figure II-9. Dogleg and 90 Degree Offset

•One pace is the distance covered each time the same foot touches the ground.

•Distances measured by paces are approximate (example in open terrain, 900 paces per kilometer [average], or example in rough terrain, 1200 paces per kilometer [average]).

(c) Use pace count in conjunction with terrain evaluation and heading to determine location. An individual's pace varies because of factors such as steep terrain, day/night travel, or injured/uninjured condition. Adjust estimation of distance traveled against these factors to get relative accuracy when using a pace count.

(4) Deliberate offset is—

(a) Used when finding a point on a linear feature (that is, road or river).

(b) Intentionally navigated to left or right of target so you know which way to turn at the linear feature.

(5) Point-to-point is same as straight line.

(a) Pick out landmarks on the heading and walk the trail of least resistance to a point.

(b) On reaching a point, establish another landmark and continue.

3. Travel Considerations

a. Pick the easiest and safest route (non-combat).

b. Maintain a realistic pace; take rest stops when needed.

c. Avoid overdressing and overheating.

d. Consider food and water requirements.

e. Take special care of feet (change socks regularly).

f. Pack equipment to prevent loss, damage, pack imbalance, and personal safety.

g. Go **around** obstacles, not over or through them.

h. Travel on trails whenever possible (non-combat).

i. Travel in forested areas if possible.

j. Avoid creek bottoms and ravines with ***NO*** escape in the event of heavy rains.

k. Consider the following for swamps, lakes, and unfordable rivers:

(1) Circumnavigate swamps, lakes, and bogs if needed.

(2) Travel downstream to find people and slower water.

(3) Travel upstream to find narrower and shallow water.

4. River Travel

River travel may be faster and save energy when hypothermia is not a factor. It may be a primary mode of travel and LOC in a tropical environment **(use with caution if evading)**.

a. Use flotation device (raft, log, bamboo, etc.).

b. Use a pole to move the raft in shallow water.

c. Use an oar in deep water.

d. Stay near inside edge of river bends (current speed is less).

e. Keep near shore.

f. Watch for the following ***DANGERS:***

(1) Snags.

(2) Sweepers (overhanging limbs and trees).

(3) Rapids (***DO NOT*** attempt to shoot the rapids).

(4) Waterfalls.

(5) Hazardous animals.

g. Consider using a flotation device when crossing rivers or large/deep streams.

5. Ice and Snow Travel

Travel should be limited to areas free of hazards.

a. ***DO NOT*** travel in—

(1) Blizzards.

(2) Bitterly cold winds.

(3) Poor visibility.

b. Obstacles to winter travel follow:

(1) Reduced daylight hours (***BE AWARE***).

(2) Deep soft snow (if movement is necessary, make snowshoes **[Figure II-10]**). Travel is easier in early morning or late afternoon near dusk when snow is frozen or crusted.

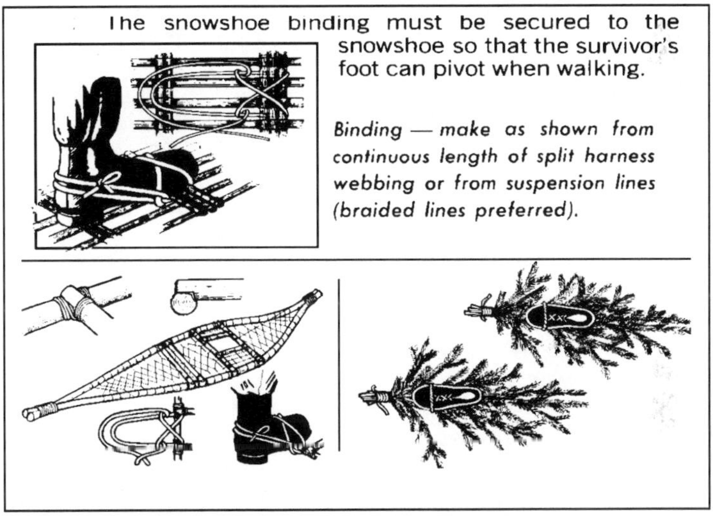

Figure II-10. Improvised Snowshoes

(3) Avalanche prone areas to avoid:

(a) Slopes 30-45 degrees or greater.

(b) Trees without uphill branches (identifies prior avalanches).

(c) Heavy snow loading on ridge tops.

(4) If caught in an avalanche, do the following:

(a) Backstroke to decrease burial depth.

(b) Move hand around face to create air pocket as moving snow slows.

(5) Frozen water crossings.

(a) Weak ice should be expected where—

- Rivers are straight.
- Objects protrude through ice.
- Snow banks extend over the ice.
- Rivers or streams come together.
- Water vapor rising indicates open or warm areas.

(b) Air pockets form when a frozen river loses volume.

(c) When crossing frozen water, distribute your weight by laying flat, belly crawling, or using snowshoes.

c. Glacier travel is hazardous and should be avoided.

6. Mountain Hazards

a. Lightning. Avoid ridge tops during thunderstorms.

b. Avalanche. Avoid areas prone to avalanches.

c. Flash floods. Avoid low areas.

7. Summer Hazards (see page II-10; paragraph 3, *Travel Considerations*, items h through k.)

(1) Dense brush.

(a) Travel on trails when possible (non-combat).

(b) Travel in forested areas if possible.

(c) Avoid creek bottoms and ravines with no escape in the event of heavy rains.

(2) Swamps, lakes, and unfordable rivers.

(a) Circumnavigate swamps, lakes, and bogs if needed.

(b) Travel downstream to find people and slower water.

(c) Travel upstream to find narrower and shallow water.

8. Dry Climates

a. ***DO NOT*** travel unless certain of reaching the destination using the water supply available.

b. Travel at dawn or dusk on hot days.

c. Follow the easiest trail possible (non-combat), avoiding—

(1) Deep sandy dune areas.

(2) Rough terrain.

d. In sand dune areas—

(1) Follow hard valley floor between dunes.

(2) Travel on the windward side of dune ridges.

e. If a sandstorm occurs—

(1) Mark your direction of travel.

(2) Sit or lie down in direction of travel.

(3) Try to get to the downwind side of natural shelter.

(4) Cover the mouth and nose with a piece of cloth.

(5) Protect the eyes.

(6) Remain stationary until the storm is over.

9. Tropical Climates

a. Travel only when it is light.

b. Avoid obstacles like thickets and swamps.

c. Part the vegetation to pass through. Avoid grabbing vegetation; it may have spines or thorns (**use gloves** if possible).

d. ***DO NOT*** climb over logs if you can go around them.

e. Find trails—

(1) Where **2** streams meet.

(2) Where a low pass goes over a range of hills.

f. While traveling trails—

(1) Watch for disturbed areas on game trails; they may indicate a pitfall or trap.

(2) Use a walking stick to probe for pitfalls or traps.

(3) ***DO NOT*** sleep on the trail.

(4) Exercise caution, the enemy uses the trails also.

10. Open Seas

a. Using currents—

(1) Deploy sea anchor **(Figure II-11)**. Sea anchor may be adjusted to make use of existing currents.

(2) Sit low in the raft.

(3) Deflate the raft slightly so it rides lower in the water.

b. Using winds—

(1) Pull in sea anchor.

(2) Inflate raft so it rides higher.

(3) Sit up in raft so body catches the wind.

(4) Construct a shade cover/sail **(Figure II-12)**. (Sail aids in making landfall.)

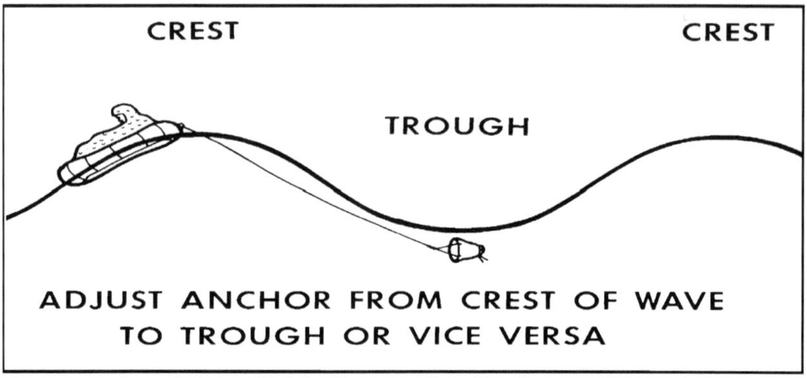

Figure II-11. Sea Anchor Deployment

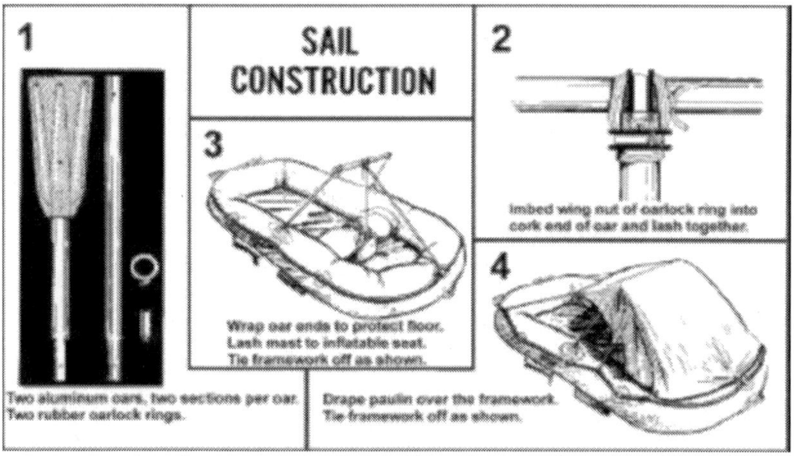

Figure II-12. Shade/Sail Construction

c. Making landfall. Indications of land are—

(1) Fixed cumulus clouds in a clear sky or in a cloudy sky where all other clouds are moving.

(2) Greenish tint in the sky **(in the tropics)**.

(3) Lighter colored reflection on clouds (open water causes dark gray reflections) (**in the arctic**).

(4) Lighter colored water (indicates shallow water).
(5) The odors and sounds.
(a) Odors from swamps and smoke.
(b) Roar of surf/bird cries coming from one direction.
(6) Directional flights of birds at dawn and at dusk.

d. Swimming ashore—
(1) Consider physical condition.
(2) Use a flotation aid.
(3) Secure all gear to body before reaching landfall.
(4) Remain in raft as long as possible.
(5) Use the sidestroke or breaststroke to conserve strength if thrown from raft.
(6) Wear footgear and at least **1** layer of clothing.
(7) Try to make landfall during the lull between the sets of waves (waves are generally in **sets** of **7**, from **smallest** to **largest**).
(8) In moderate surf.
(a) Swim forward on the back of a wave.
(b) Make a shallow dive just before the wave breaks to end the ride.
(9) In high surf.
(a) Swim shoreward in the trough between waves.
(b) When the seaward wave approaches, face it and submerge.
(c) After it passes, work shoreward in the next trough.
(10) If caught in the undertow of a large wave—
(a) Remain calm and swim to the surface.
(b) Lie as close to the surface as possible.
(c) Parallel shoreline and attempt landfall at a point further down shore.
(11) Select a landing point.
(a) Avoid places where waves explode upon rocks.
(b) Find a place where waves smoothly rush onto the rocks.
(12) After selecting a landing site—
(a) Face shoreward.
(b) Assume a sitting position with feet 2 or 3 feet lower than head to absorb the shock of hitting submerged objects.

e. Rafting ashore—

(1) Select landing point carefully.

(2) Use caution landing when the sun is low and straight in front of you causing poor visibility.

(3) Land on the lee (downwind) side of islands or point of land if possible.

(4) Head for gaps in the surf line.

(5) Penetrate surf by—

(a) Taking down most shade/sails.

(b) Using paddles to maintain control.

(c) Deploying a sea anchor for stability.

CAUTION: ***DO NOT*** deploy a sea anchor if traveling through coral.

f. Making sea ice landings on large stable ice flows. Icebergs, small flows, and disintegrating flows are dangerous **(ice can cut a raft)**.

(1) Use paddles to avoid sharp edges.

(2) Store raft away from the ice edge.

(3) Keep raft inflated and ready for use.

(4) Weight down/secure raft so it does not blow away.

Chapter III

RADIO COMMUNICATIONS AND SIGNALING

Inventory and review the operating instructions of all communications and signaling equipment.

1. Radio Communications (Voice and Data)

a. Non-combat.

(1) Ensure locator beacon is operational.

(2) Follow standing plans for on/off operations to conserve battery use.

b. Combat.

(1) Turn off locator beacon.

(2) Keep it with you to supplement radio communications.

(3) Follow plans/orders for on/off operations.

c. Make initial contact as soon as possible or as directed in applicable plans/orders.

d. If no immediate contact, then as directed in applicable plans/orders.

e. Locate spare radio and batteries (keep warm and dry).

f. Transmissions.

(1) Use concealment sites (combat) that optimize line of site (LOS).

(2) Face recovery asset.

(3) Keep antenna perpendicular to intended receiver **(Figure III-1)**.

(4) ***DO NOT*** ground antenna (that is finger on antenna or attaching bolt, space blanket, vegetation, etc.).

(5) Keep transmissions short (3-5 seconds maximum). Use data burst if available.

(6) Move after each transmission (***ONLY*** in combat, if possible).

(7) If transmitting in the blind, ensure a clear LOS towards the equator.

(8) Use terrain masking to hinder enemy direction finding.

g. Listening (use reception times in applicable plans/orders or as directed by recovery forces).

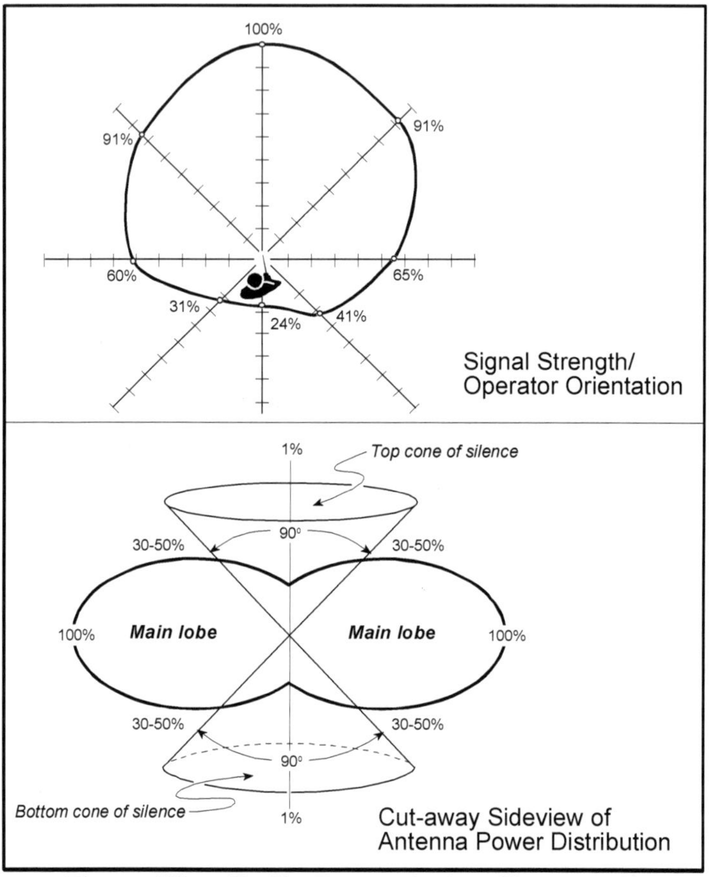

Figure III-1. Radio Transmission Characteristics

2. Signaling

a. Pyrotechnic signals.

(1) Prepare early (weather permitting).

(2) Use as directed in applicable plans/orders or as directed by recovery forces.

(3) Extend over raft's edge before activating.

b. Signal mirror **(Figure III-2)**.
  (1) Use as directed by recovery forces.
  (2) If no radio, use only with confirmed friendly forces.
  (3) Cover when not in use.

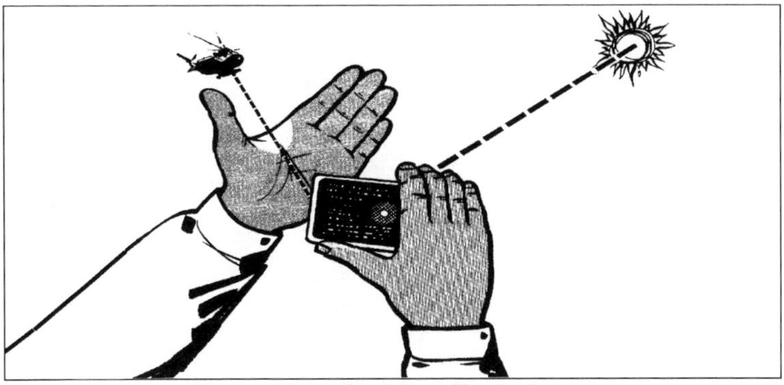

Figure III-2. Sighting Techniques

Note: Make a mirror from any shiny metal or glass.

c. Strobe/IR lights.
  (1) Prepare early, consider filters and shields.
  (2) Use as directed by recovery forces.
  (3) Conserve battery life.

Note: Produces one residual flash when turned off.

d. Pattern signals (use as directed in applicable plans/orders).
  (1) Materials:
    (a) Manmade (space blanket, signal paulin, parachute).
    (b) Natural use materials that contrast the color and/or texture of the signaling area (rocks, brush, branches, stomped grass).
  (2) Location.
    (a) Maximize visibility from above.
    (b) Provide concealment from ground observation.
  (3) Size (large as possible) and ratio **(Figure III-3)**.

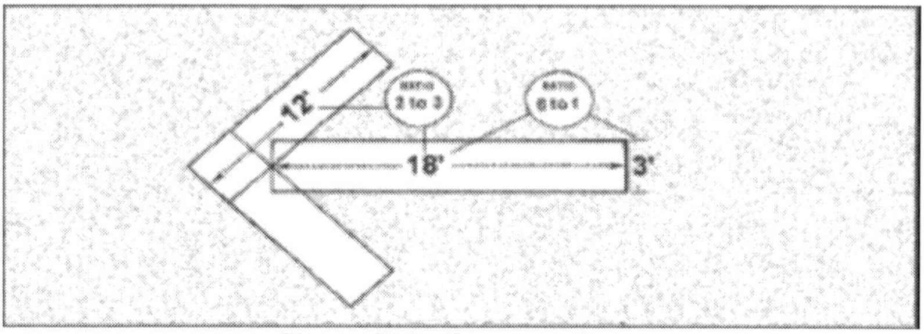

Figure III-3. Size and Ratio

(4) Shape (maintain straight lines and sharp corners).
(5) Contrast (use color and shadows).
(6) Pattern signals **(Figure III-4)**.

NO.	MESSAGE	CODE SYMBOL
1	REQUIRE ASSISTANCE	V
2	REQUIRE MEDICAL ASSISTANCE	X
3	NO or NEGATIVE	N
4	YES or AFFIRMATIVE	Y
5	PROCEEDING IN THIS DIRECTION	↑

Figure III-4. Signal Key

e. Sea dye marker.
   (1) ***DO NOT*** waste in rough seas or fast moving water.
   (2) Conserve unused dye by rewrapping.
   (3) May be used to color snow.

f. Non-combat considerations:

(1) Use a fire at night.

(2) Use smoke for day (tires or petroleum products for dark smoke and green vegetation for light smoke). **(Figure III-5)**

(3) Use signal mirror to sweep horizon.

(4) Use audio signals (that is, voice, whistle, and weapons fire).

Figure III-5. Smoke Generator

Chapter IV
RECOVERY

1. Responsibilities

a. Establish radio contact with recovery forces (if possible).

b. Maintain communication with recovery forces until recovered.

c. Be prepared to authenticate as directed in applicable plans/orders.

d. Follow recovery force instructions, be prepared to report—

(1) Enemy activity in the recovery area.

(2) Recovery site characteristics (slope, obstacles, size, etc.).

(3) Number in party/medical situation.

(4) Signal devices available.

e. If no radio, a ground-to-air signal may be your only means to effect recovery.

2. Site Selection

a. Locate area for landing pick-up, if practical (approximately 150 feet diameter, free of obstructions, flat and level).

b. Assess evidence of human activity at/near the site (in combat).

c. Locate several concealment sites around area (in combat).

d. Plan several tactical entry and exit routes (in combat).

3. Site Preparation

a. Pack and secure all equipment.

b. Prepare signaling devices (use as directed or as briefed).

c. Mentally review recovery methods (aircraft, ground, boat, etc.).

4. Recovery Procedures

a. Assist recovery force in identifying your position.

b. Stay concealed until recovery is imminent (in combat).

c. For a landing/ground recovery—

(1) Assume a non-threatening posture.

(2) Secure weapons and avoid quick movement.

(3) ***DO NOT*** approach recovery vehicle until instructed.

(4) Beware of rotors/propellers when approaching recovery vehicle, especially on sloping or uneven terrain. Secure loose equipment that could be caught in rotors/propellers.

d. For **hoist** recovery devices **(Figures IV-1 and IV-2)**—

(1) Use eye protection, if available (glasses or helmet visor).

(2) Allow metal on device to contact the surface before touching to avoid injury from static discharge.

(3) Sit or kneel for stability while donning device.

(4) Put safety strap under armpits.

(5) Ensure cable is in front of you.

(6) Keep hands clear of all hardware and connectors.

(7) ***DO NOT*** become entangled in cable.

(8) Use a thumbs up, vigorous cable shake, or radio call to signal you are ready.

(9) Drag feet on the ground to decrease oscillation.

(10) ***DO NOT*** assist during hoist or when pulled into the rescue vehicle. Follow crewmember instructions.

e. For **nonhoist** recovery (rope or unfamiliar equipment)—

(1) Create a ***"fixed loop"*** big enough to place under armpits **(Figure IV-3)**.

(2) Follow the procedures in "**d**" above.

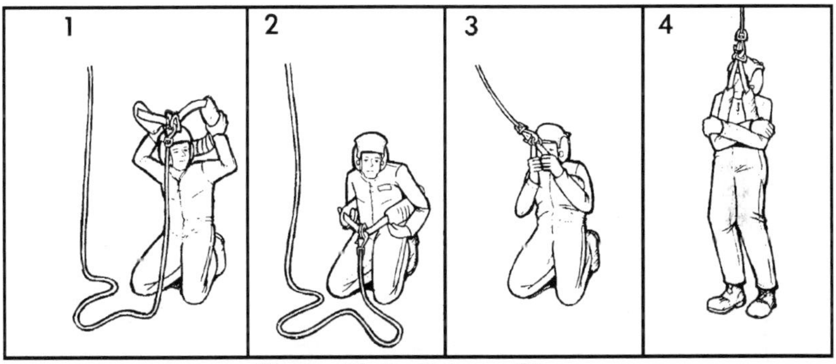

Figure IV–1. Rescue Strap

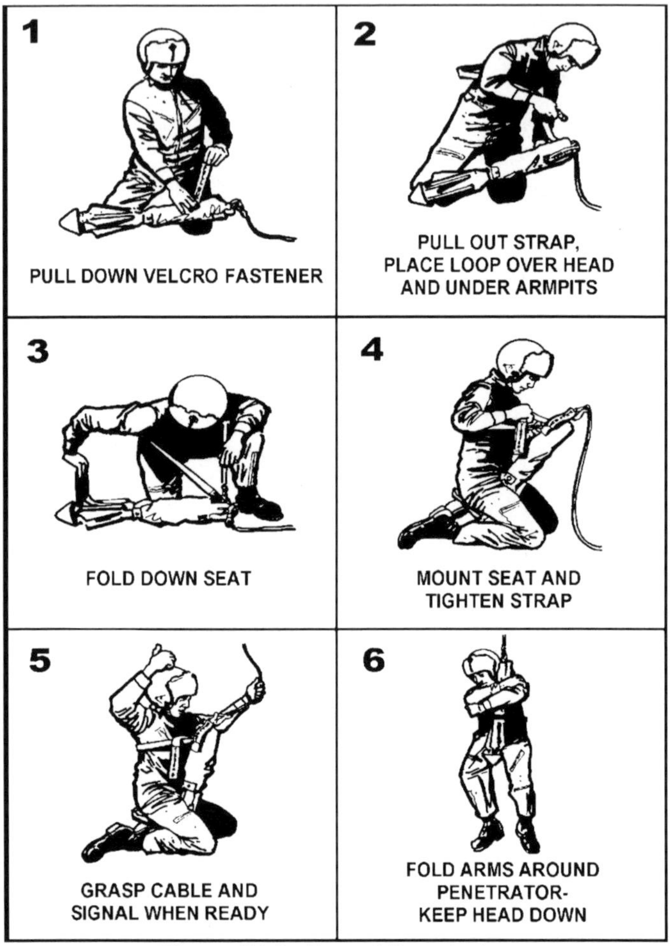

Figure IV-2. Forest Penetrator

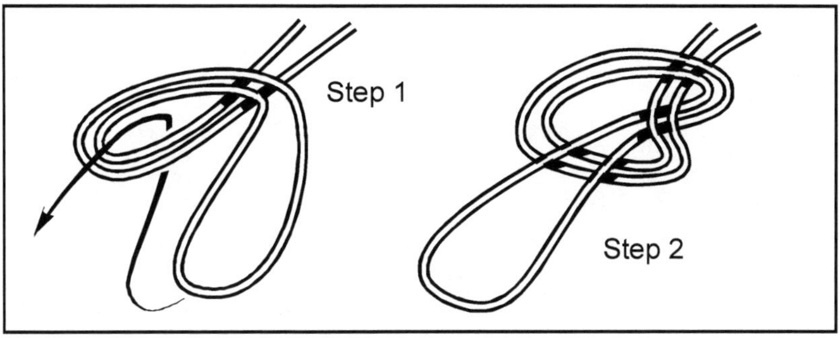

Figure IV–3. Fixed Loop

Chapter V

MEDICAL

WARNING: These emergency medical procedures are for survival situations. Obtain professional medical treatment as soon as possible.

1. Immediate First Aid Actions

Remember the *ABCs* of Emergency Care:
Airway Breathing Circulation

a. Determine responsiveness as follows:
   (1) If unconscious, arouse by shaking gently and shouting.
   (2) If no response—
      (a) Keep head and neck aligned with body.
      (b) Roll victims onto their backs.
      (c) Open the airway by lifting the chin **(Figure V-1)**.
      (d) Look, listen, and feel for air exchange.

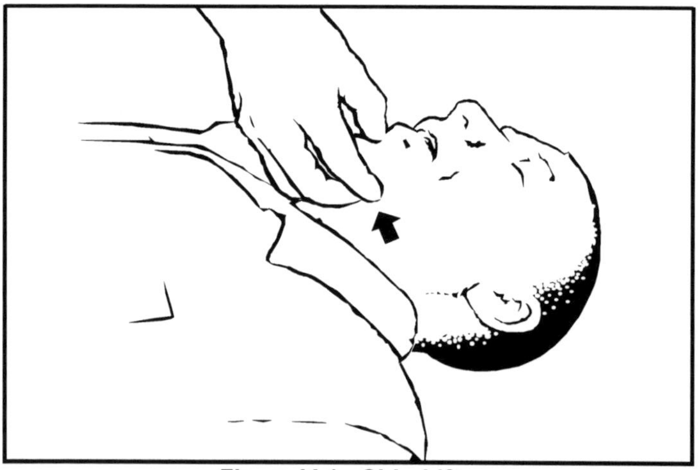

Figure V-1. Chin Lift

(3) If victim is not breathing—
(a) Check for a clear airway; remove any blockage.
(b) Cover victim's mouth with your own.
(c) Pinch victim's nostrils closed.
(d) Fill victim's lungs with **2** slow breaths.
(e) If breaths are blocked, reposition airway; try again.
(f) If breaths still blocked, give **5** abdominal thrusts:
•Straddle the victim.
•Place a fist between breastbone and belly button.
•Thrust upward to expel air from stomach.
(g) Sweep with finger to clear mouth.
(h) Try **2** slow breaths again.
(i) If the airway is still blocked, continue **(c)** through **(f)** until successful or exhausted.
(j) With open airway, start mouth to mouth breathing:
•Give **1** breath every 5 seconds.
•Check for chest rise each time.
(4) If victim is unconscious, but breathing—
(a) Keep head and neck aligned with body.
(b) Roll victim on side (drains the mouth and prevents the tongue from blocking airway).
(5) If breathing difficulty is caused by chest trauma, refer to **page V-7, paragraph 1d, *Treat Chest Injuries*.**

CAUTION: ***DO NOT*** remove an impaled object unless it interferes with the airway. You may cause more tissue damage and increase bleeding. For travel, you may shorten and secure the object.

b. Control bleeding as follows:
(1) Apply a pressure dressing **(Figure V-2)**.
(2) If ***STILL*** bleeding—
(a) Use direct pressure over the wound.
(b) Elevate the wounded area above the heart.

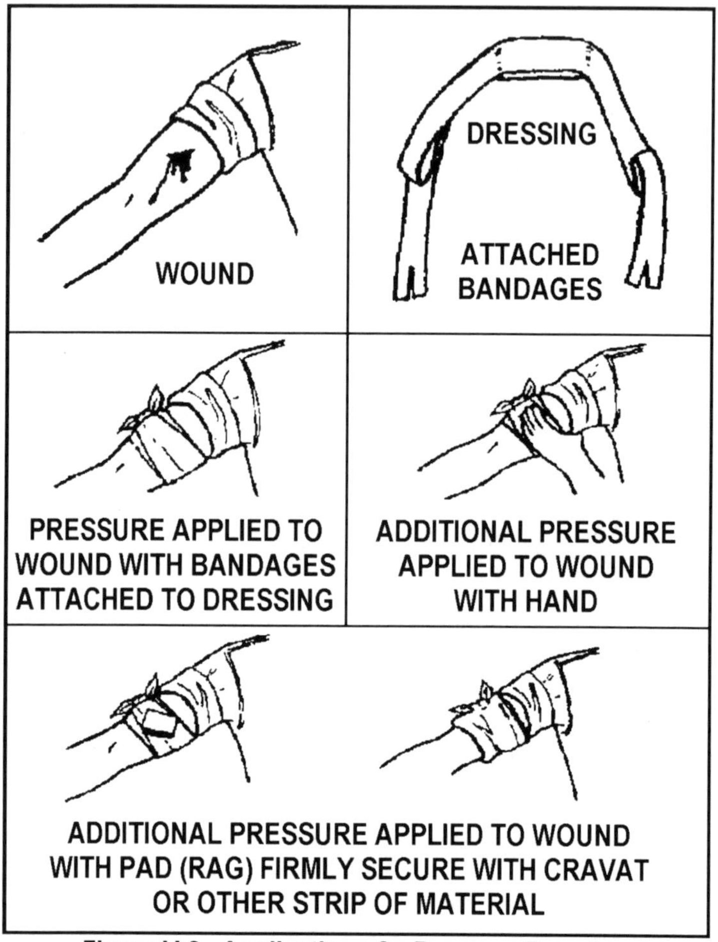

Figure V-2. Application of a Pressure Dressing

(3) If ***STILL*** bleeding—

(a) Use a pressure point between the injury and the heart **(Figure V-3)**.

(b) Maintain pressure for 6 to 10 minutes before checking to see if bleeding has stopped.

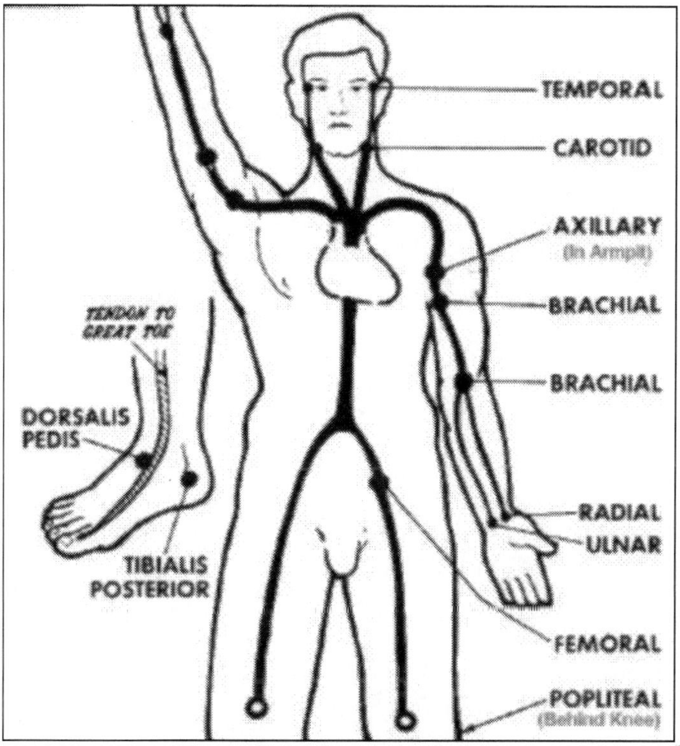

Figure V-3. Pressure Points

(4) If a limb wound is ***STILL*** bleeding—

CAUTION: Use of a tourniquet is a ***LAST RESORT*** measure. Use ***ONLY*** when severe, uncontrolled bleeding will cause loss of life. Recognize that long-term use of a tourniquet may cause loss of limb.

(a) Apply tourniquet (TK) band just above bleeding site on limb. A band at least 3 inches (7.5 cm) or wider is best.

(b) Follow steps illustrated in **Figure V-4**.

(c) Use a stick at least 6 inches (15 cm) long.

(d) Tighten only enough to stop arterial bleeding.

(e) Mark a ***TK*** on the forehead with the time applied.

(f) ***DO NOT*** cover the tourniquet.

CAUTION: The following directions apply ***ONLY*** in survival situations where rescue is ***UNLIKELY*** and ***NO*** medical aid is available.

(g) If rescue or medical aid is not available for over 2 hours, an attempt to ***SLOWLY*** loosen the tourniquet may be made 20 minutes after application. Before loosening—

- Ensure pressure dressing is in place.
- Ensure bleeding has stopped
- Loosen tourniquet ***SLOWLY*** to restore circulation.
- Leave loosened tourniquet in position in case bleeding resumes.

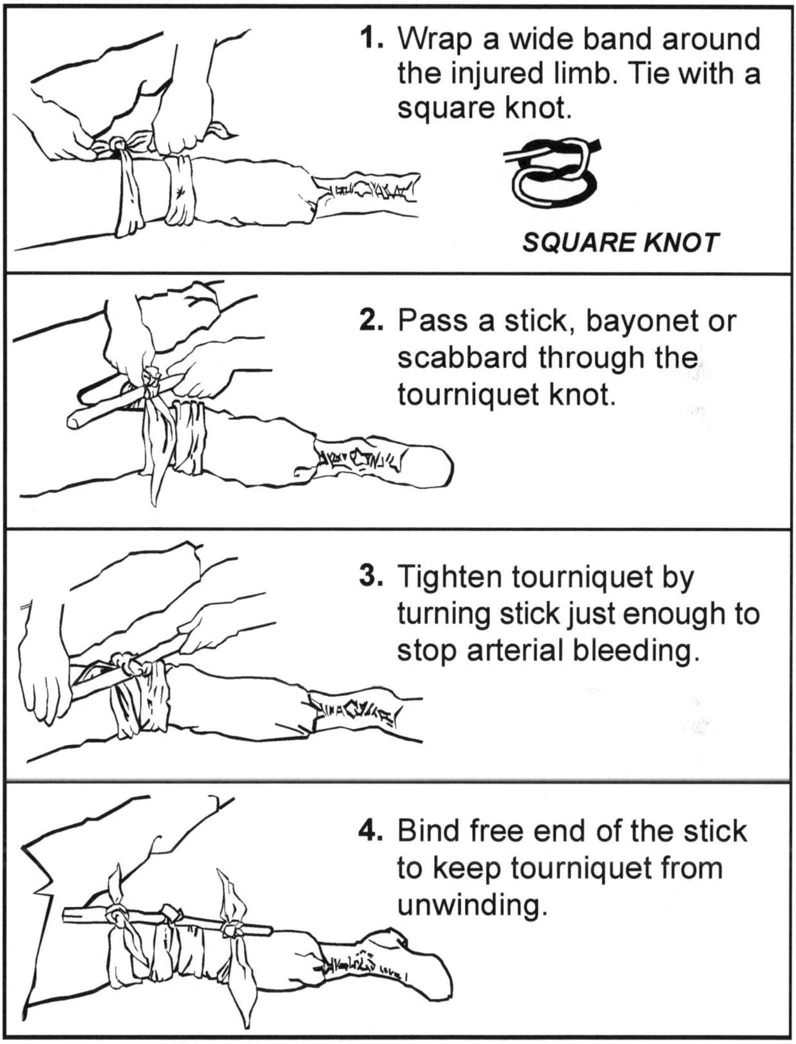

Figure V-4. Application of a Tourniquet

c. Treat shock. (Shock is difficult to identify or treat under field conditions. It may be present with or without visible injury.)

(1) Identify by one or more of the following:

(a) Pale, cool, and sweaty skin.

(b) Fast breathing and a weak, fast pulse.

(c) Anxiety or mental confusion.

(d) Decreased urine output.

(2) Maintain circulation.

(3) Treat underlying injury.

(4) Maintain normal body temperature.

(a) Remove wet clothing.

(b) Give warm fluids.

•***DO NOT*** give fluids to an unconscious victim.

•***DO NOT*** give fluids if they cause victim to gag.

(c) Insulate from ground.

(d) Shelter from the elements.

(5) Place conscious victim on back.

(6) Place very weak or unconscious victim on side, this will—

(a) Allow mouth to drain.

(b) Prevent tongue from blocking airway.

d. Treat chest injuries.

(1) Sucking chest wound. This occurs when chest wall is penetrated; may cause victim to gasp for breath; may cause sucking sound; may create bloody froth as air escapes the chest.

(a) ***Immediately*** seal wound with hand or airtight material.

(b) Tape airtight material over wound on ***3 sides only*** **(Figure V-5)** to allow air to escape from the wound but not to enter.

(c) Monitor breathing and check dressing.

(d) Lift untapped side of dressing as victim ***exhales*** to allow trapped air to escape, as necessary.

(2) Flail chest. Results from blunt trauma when ***3*** or ***more*** ribs are broken in ***2*** or more places. The flail segment is the broken area that moves in a direction opposite to the rest of chest during breathing.

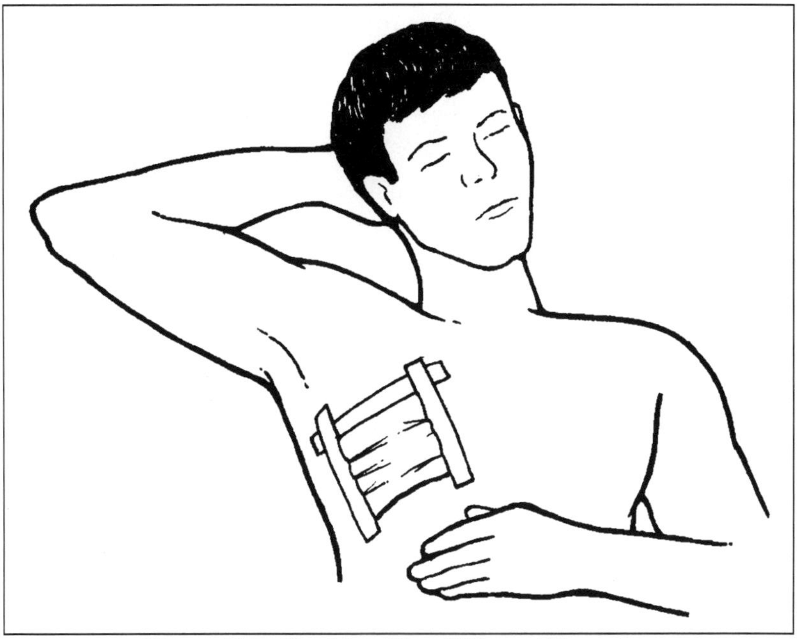

Figure V-5. Sucking Chest Wound Dressing

(a) Stabilize the flail segment as follows:
•Place rolled-up clothing or bulky pad over site.
•Tape pad to site
•***DO NOT*** wrap tape around chest.

(b) Have victim keep segment still with hand pressure.

(c) Roll victim onto side of flail segment injury (as other injuries allow).

(3) Fractured ribs.

(a) Encourage deep breathing (painful, but necessary to prevent the possible development of pneumonia).

(b) ***DO NOT*** constrict breathing by taping ribs.

e. Treat fractures, sprains, and dislocations.

(1) Control bleeding.

(2) Remove watches, jewelry, and constrictive clothing.

(3) If fracture penetrates the skin—

(a) Clean wound by gentle irrigation with water.

(b) Apply dressing over wound.

(4) Position limb as normally as possible.

(5) Splint in position found (if ***unable*** to straighten limb).

(6) Improvise a splint with available materials:

(a) Sticks or straight, stiff materials from equipment.

(b) Body parts (for example, opposite leg, arm-to-chest).

(7) Attach with strips of cloth, parachute cord, etc.

(8) Keep the fractured bones from moving by immobilizing the joints on both sides of the fracture. If fracture is in a joint, immobilize the bones on both sides of the joint.

CAUTION: Splint fingers in a slightly flexed position, ***NOT*** in straight position. Hand should look like it is grasping an apple.

(9) Use ***RICES*** treatment for 72 hours.

(a) **R**est.

(b) **I**ce.

(c) **C**ompression.

(d) **E**levation.

(e) **S**tabilization.

(10) Apply cold to acute injuries.

(11) Use 15 to 20 minute periods of cold application.

(a) ***DO NOT*** use continuous cold therapy.

(b) Repeat 3 to 4 times per day.

(c) Avoid cooling that can cause frostbite or hypothermia.

(12) Wrap with a compression bandage after cold therapy.

(13) Elevate injured area above heart level to reduce swelling.

(14) Check periodically for a pulse beyond the injury site.

(15) Loosen bandage or reapply splint if no pulse is felt or if swelling occurs because bandage is too tight.

2. Common Injuries and Illnesses

a. Burns.

(1) Cool the burned area with water.

(a) Use immersion or cool compresses.

(b) Avoid aggressive cooling with ice or frigid water.

(2) Remove watches, jewelry, constrictive clothing.

(3) ***DO NOT*** remove embedded, charred material that will cause burned areas to bleed.

(4) Cover with sterile dressings.

(5) ***DO NOT*** use lotion or grease.

(6) Avoid moving or rubbing the burned part.

(7) Drink ***extra*** water to compensate for increased fluid loss from burns. (Add ***1/4 teaspoon*** of ***salt*** [if available] to ***each quart*** of ***water***.)

(8) Change dressings when soaked or dirty.

b. Eye injuries.

(1) Sun/snow blindness (gritty, burning sensation, and possible reduction in vision caused by sun exposure).

(a) Prevent with improvised goggles. **(See Chapter VI, page VI-3, Figure VI-2.)**

(b) Treat by patching affected eye(s).

- Check after 12 hours.
- Replace patch for another 12 hours if not healed.

(c) Use cool compresses to reduce pain.

(2) Foreign body in eye.

(a) Irrigate with clean water from the ***inside*** to the ***outside*** corner of the eye.

(b) If foreign body is not removed by irrigation, improvise a small swab. Moisten and wipe gently over the affected area.

(c) If foreign body is ***STILL*** not removed, patch eye for 24 hours and then reattempt removal using steps **(a)** and **(b)**.

c. Heat injury.

(1) Heat cramps (cramps in legs or abdomen).

(a) Rest.

(b) Drink water. Add ***1/4 teaspoon*** of salt ***per quart***.

(2) Heat exhaustion (pale, sweating, moist, cool skin).

(a) Rest in shade.

(b) Drink water.

(c) Protect from further heat exposure.

(3) Heat stroke (victim disoriented or unconscious, skin is hot and flushed [sweating **may** or **may not** occur], fast pulse).

CAUTION: Handle heat stroke victim gently. Shock, seizures, and cardiac arrest can occur.

(a) Cool as rapidly as possible (saturate clothing with water and fan the victim). Remember to cool the groin and armpit areas. (Avoid overcooling.)

(b) Maintain airway, breathing, and circulation.

d. Cold injuries:

(1) Frostnip and frostbite—

(a) Are progressive injuries.

•Ears, nose, fingers, and toes are affected first.

•Areas will feel cold and may tingle leading to—

••Numbness that progresses to—

•••Waxy appearance with stiff skin that cannot glide freely over a joint.

(b) Frostnipped areas rewarm with body heat. If body heat ***WILL NOT*** rewarm area in 15 to 20 minutes, then frostbite is present.

(c) Frostbitten areas are deeply frozen and require medical treatment.

CAUTION: In frostbite, repeated freezing and thawing causes severe pain and increases damage to the tissue. ***DO NOT*** rub frozen tissue. ***DO NOT*** thaw frozen tissue.

(2) Hypothermia—

(a) Is a progressive injury.

•Intense shivering with impaired ability to perform complex tasks leads to—

••Violent shivering, difficulty speaking, sluggish thinking go to—

•••Muscular rigidity with blue, puffy skin; jerky movements go to—

••••Coma, respiratory and cardiac failure.

(b) Protect victim from the environment as follows:

•Remove wet clothing.

•Put on dry clothing (if available).

•Prevent further heat loss.

••Cover top of head.

••Insulate from above and below.

•Warm with blankets, sleeping bags, or shelter.

•Warm central areas before extremities.

••Place heat packs in groin, armpits, and around neck.

••Avoid causing burns to skin.

> **CAUTION**: Handle hypothermia victim gently. Avoid overly rapid rewarming which may cause cardiac arrest. Rewarming of victim with skin-to-skin contact by volunteer(s) inside of a sleeping bag is a survival technique but can cause internal temperatures of all to drop.

e. Skin tissue damage.

(1) Immersion injuries. Skin becomes wrinkled as in ***dishpan hands***.

(a) Avoid walking on affected feet.
(b) Pat dry; ***DO NOT*** rub. Skin tissue will be sensitive.
(c) Dry socks and shoes. Keep feet protected.
(d) Loosen boots, cuffs, etc., to improve circulation.
(e) Keep area dry, warm, and open to air.
(f) ***DO NOT*** apply creams or ointments.

(2) Saltwater sores.

(a) Change body positions frequently.
(b) Keep sores dry.
(c) Use antiseptic (if available).
(d) ***DO NOT*** open or squeeze sores.

f. Snakebite.

> **CAUTION:** This snakebite treatment recommendation is for situations where medical aid and specialized equipment are not available.

(1) Nonpoisonous. Clean and bandage wound.

(2) Poisonous.

(a) Remove constricting items.
(b) Minimize activity.
(c) ***DO NOT*** cut the bite site; ***DO NOT*** use your mouth to create suction.
(d) Clean bite with soap and water; cover with a dressing.

(e) Overwrap the bite site with a tight (elastic) bandage **(Figure V-6)**. The intent is to slow capillary and venous blood flow but not arterial flow. Check for pulse below the overwrap.

(f) Splint bitten extremity to prevent motion.

(g) Treat for shock **(page V-7, paragraph 1c)**.

(h) Position extremity below level of heart.

(i) Construct shelter if necessary (let the victim rest).

(j) For conscious victims, force fluids.

g. Marine life.

(1) Stings.

(a) Flush wound with salt water (fresh water stimulates toxin release).

(b) Remove jewelry and watches.

(c) Remove tentacles and gently scrape or shave skin.

(d) Apply a steroid cream (if available).

(e) ***DO NOT*** rub area with sand.

(f) Treat for shock; artificial respiration may be required **(page V-1, paragraph 1a)**.

(g) ***DO NOT*** use urine to flush or treat wounds.

(2) Punctures.

(a) Immerse affected part in hot water or apply hot compresses for 30-60 minutes (as hot as victim can tolerate).

(b) Cover with clean dressing.

(c) Treat for shock as needed.

h. Skin irritants (includes poison oak and poison ivy).

(1) Wash with large amounts of water. Use soap (if available).

(2) Keep covered to prevent scratching.

i. Infection.

(1) Keep wound clean.

(2) Use iodine tablet solution or diluted betadine to prevent or treat infection.

(3) Change bandages as needed.

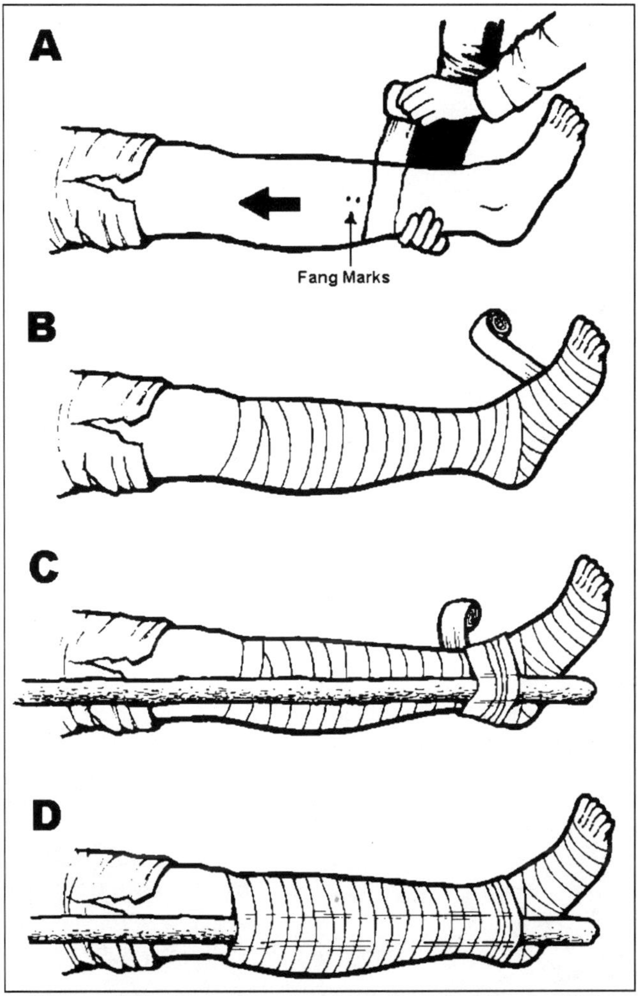

Figure V-6. Compression Bandage for Snake Bite

j. Dysentery and diarrhea.

(1) Drink ***extra*** water.

(2) Use a liquid diet.

(3) Eat charcoal. Make a paste by mixing fine charcoal particles with water. (It may relieve symptoms by absorbing toxins.)

k. Constipation (can be expected in survival situations).

(1) ***DO NOT*** take laxatives.

(2) Exercise.

(3) Drink **extra** water.

3. Plant Medicine

a. Tannin.

(1) Medical uses. Burns, diarrhea, dysentery, skin problems, and parasites. Tannin solution prevents infection and aids healing.

(2) Sources. Found in the outer bark of all trees, acorns, banana plants, common plantain, strawberry leaves, and blackberry stems.

(3) Preparation.

(a) Place crushed outer bark, acorns, or leaves in water.

(b) Leach out the tannin by soaking or boiling.

- Increase tannin content by longer soaking time.
- Replace depleted material with fresh bark/plants.

(4) Treatments.

(a) Burns.

- Moisten bandage with cooled tannin tea.
- Apply compress to burned area.
- Pour cooled tea on burned areas to ease pain.

(b) Diarrhea, dysentery, and worms. Drink strong tea solution (may promote voiding of worms).

(c) Skin problems (dry rashes and fungal infections). Apply cool compresses or soak affected part to relieve itching and promote healing.

(d) Lice and insect bites. Wash affected areas with tea to ease itching.

b. Salicin/salicylic acid.

(1) Medical uses. Aches, colds, fever, inflammation, pain, sprains, and sore throat (aspirin-like qualities).

(2) Sources. Willow and aspen trees **(Figure V-7)**.

(3) Preparation.

(a) Gather twigs, buds, or cambium layer (soft, moist layer between the outer bark and the wood) of willow or aspen.

(b) Prepare tea as described in paragraph **3a(3)**.

(c) Make poultice.

•Crush the plant or stems.

•Make a pulpy mass.

(4) Treatments.

(a) Chew on twigs, buds, or cambium for symptom relief.

(b) Drink tea for colds and sore throat.

(c) Use warm, moist poultice for aches and sprains.

•Apply pulpy mass over injury.

•Hold in place with a dressing.

c. Common plantain.

(1) Medical uses. Itching, wounds, abrasions, stings, diarrhea, and dysentery.

(2) Source. There are over 200 plantain species with similar medicinal properties. The common plantain is shown in **Figure V-7**.

(3) Preparation.

(a) Brew tea from seeds.

(b) Brew tea from leaves.

(c) Make poultice of leaves.

(4) Treatments.

(a) Drink tea made from seeds for diarrhea or dysentery.

(b) Drink tea made from leaves for vitamin and minerals.

(c) Use poultice to treat cuts, sores, burns, and stings

d. Papain.

(1) Medical uses. Digestive aid, meat tenderizer, and a food source.

(2) Source. Fruit of the papaya tree **(Figure V-7)**.

(3) Preparation.

(a) Make cuts in **unripe** fruit.

(b) Gather milky white sap for its papain content.

(c) Avoid getting sap in eyes or wounds.

(4) Treatments.

(a) Use sap to tenderize tough meat.

(b) Eat **ripe** fruit for food, vitamins, and minerals.

e. Common Cattail.

(1) Medical uses. Wounds, sores, boils, inflammations, burns, and an excellent food source.

(2) Source. Cattail plant found in marshes **(Figure V-7)**.

(3) Preparation.

(a) Pound roots into a pulpy mass for a poultice.

(b) Cook and eat green bloom spikes.

(c) Collect yellow pollen for flour substitute.

(d) Peel and eat tender shoots (raw or cooked).

(4) Treatments.

(a) Apply poultice to affected area.

(b) Use plant for food, vitamins, and minerals.

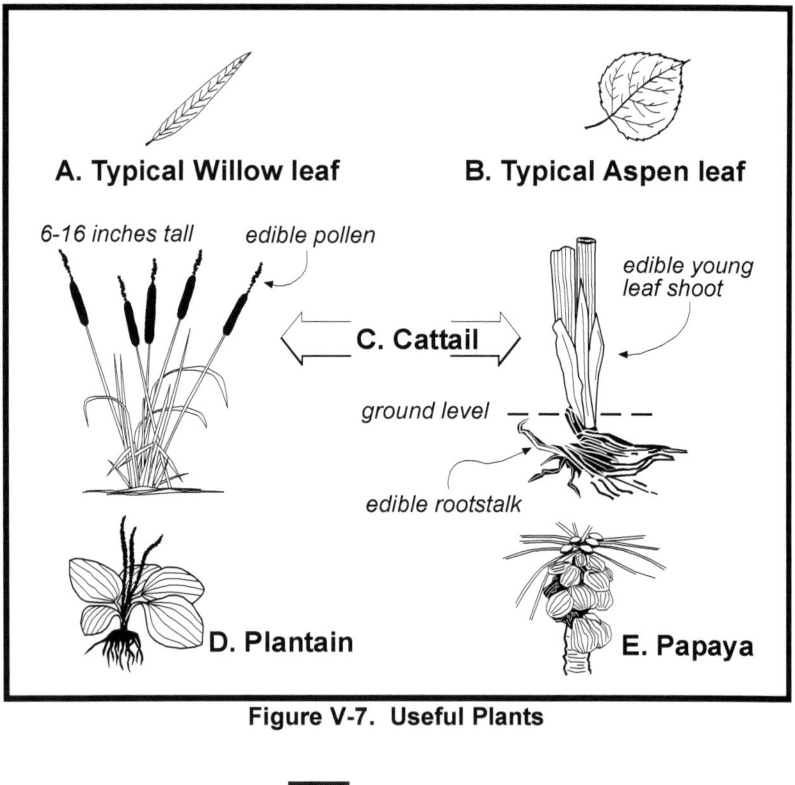

Figure V-7. Useful Plants

4. Health and Hygiene

a. Stay clean (daily regimen).

(1) Minimize infection by washing. (Use white ashes, sand, or loamy soil as soap substitutes.)

(2) Comb and clean debris from hair.

(3) Cleanse mouth and brush teeth.

(a) Use hardwood twig as toothbrush (fray it by chewing on one end then use as brush).

(b) Use single strand of an inner core string from parachute cord for dental floss.

(c) Use clean finger to stimulate gum tissues by rubbing.

(d) Gargle with salt water to help prevent sore throat and aid in cleaning teeth and gums.

(4) Clean and protect feet.

(a) Change and wash socks

(b) Wash, dry, and massage.

(c) Check frequently for blisters and red areas.

(d) Use adhesive tape/mole skin to prevent damage.

b. Exercise daily.

c. Prevent and control parasites.

(1) Check body for lice, fleas, ticks, etc.

(a) Check body regularly.

(b) Pick off insects and eggs (***DO NOT*** crush).

(2) Wash clothing and use repellents.

(3) Use smoke to fumigate clothing and equipment.

5. Rules for Avoiding Illness

a. Purify all water obtained from natural sources by using iodine tablets, bleach, or boiling for 5 minutes.

b. Locate latrines 200 feet from water and away from shelter.

c. Wash hands before preparing food or water.

d. Clean all eating utensils after each meal.

e. Prevent insect bites by using repellent, netting, and clothing.

f. Dry wet clothing as soon as possible.

g. Eat varied diet.

h. Try to get 7-8 hours sleep per day.

Chapter VI

PERSONAL PROTECTION

1. Priorities

a. Evaluate available resources and situation, then accomplish individual tasks accordingly.

b. First 24 hours in order of situational needs—

(1) Construct survival shelter according to selection criteria.

(2) Procure water.

(3) Establish multiple survival signals.

(4) Build Fire.

c. Second 24 hours in order of situational needs—

(1) Construct necessary tools and weapons.

(2) Procure food.

2. Care and Use of Clothing

a. Never discard clothing.

b. Wear loose and layered clothing.

(1) Tight clothing restricts blood flow regulating body temperature.

(2) Layers create more dead air space.

c. Keep entire body covered to prevent sunburn and dehydration in hot climates. When fully clothed, the majority of body heat escapes through the head and neck areas.

d. Avoid overheating.

(1) Remove layers of clothing before strenuous activities.

(2) Use a hat to regulate body heat.

(3) Wear a hat when in direct sunlight (in hot environment).

e. Dampen clothing when on the ocean in hot weather.

(1) Use salt water, ***NOT*** drinking water.

(2) Dry clothing before dark to prevent hypothermia.

f. Keep clothing dry to maintain its insulation qualities (dry damp clothing in the sun or by a fire).

g. If you fall into the water in the winter—

(1) Build fire.

(2) Remove wet clothing and rewarm by fire.

(3) Finish drying clothing by fire.

h. If no fire is available—
   (1) Remove clothing and get into sleeping bag (if available).
   (2) Allow wet clothes to freeze.
   (3) Break ice out of clothing.

i. Keep clothing clean (dirt reduces its insulation qualities). Examine clothing frequently for damage.
   (1) ***DO NOT*** sit or lie directly on the ground.
   (2) Wash clothing whenever possible.
   (3) Repair when necessary by using—
      (a) Needle and thread.
      (b) Safety pins.
      (c) Tape.

j. Improvised foot protection **(Figure VI-1)**.
   (1) Cut **2** to **4** layers of cloth into a 30-inch square.
   (2) Fold into a triangle.
   (3) Center foot on triangle with toes toward corner.

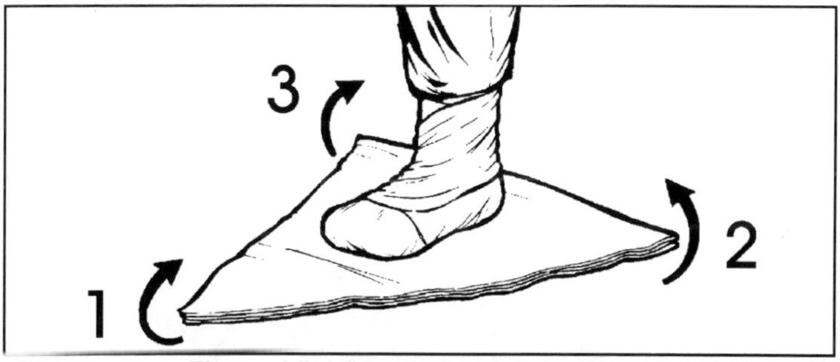

Figure VI-1 Improvised Foot Wear

   (4) Fold front over the toes.
   (5) Fold side corners, one at a time, over the instep.
   (6) Secure by rope, vines, tape, etc., or tuck into other layers of material.

3. Other Protective Equipment

a. Sleeping bag.
   (1) Fluff before use, ***especially*** at foot of bag.
   (2) Air and dry daily to remove body moisture.

(3) Improvise with available material, dry grass, leaves, dry moss, etc.

b. Sun and snow goggles **(Figure VI-2)**.

(1) Wear in bright sun or snow conditions.

(2) Improvise by cutting small horizontal slits in webbing, bark, or similar materials.

Figure VI-2. Sun and Snow Goggles

c. Gaiters **(Figure VI-3)**. Used to protect from sand, snow, insects, and scratches (wrap material around lower leg and top of boots).

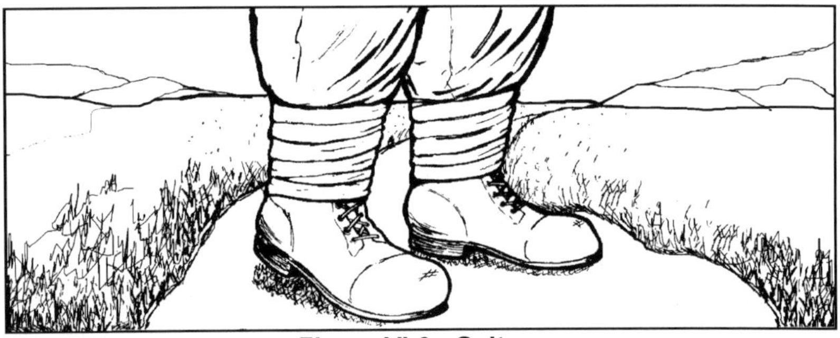

Figure VI-3. Gaiters

4. Shelters

Evasion considerations apply.

a. Site selection.

(1) Near signal and recovery site.

(2) Available food and water.

(3) Avoid natural hazards:

(a) Dead standing trees.

(b) Drainage and dry river beds except in combat areas.

(c) Avalanche areas.

(4) Location large and level enough to lie down in.

b. Types.

(1) Immediate shelters. Find shelter needing minimal improvements **(Figure VI-4)**.

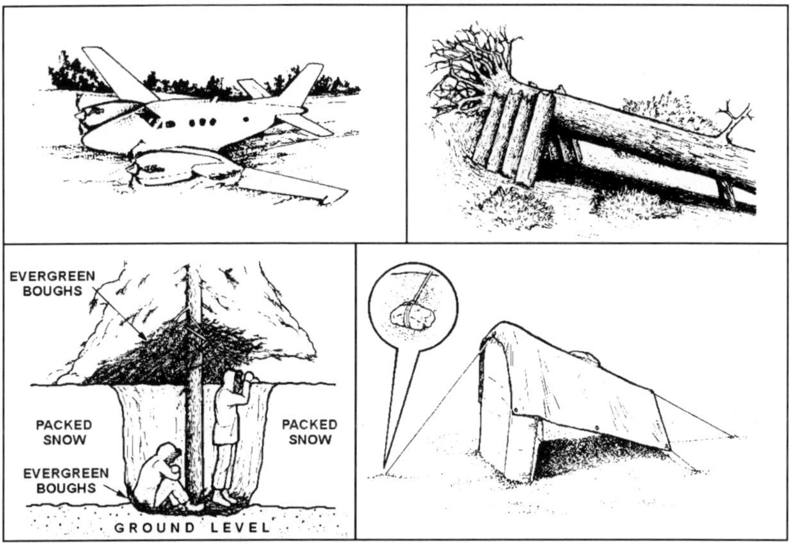

Figure VI-4. Immediate Shelters

(2) General shelter. Temperate climates require any shelter that gives protection from wind and rain.

(3) Thermal A Frame, Snow Trench, Snow Cave. **(Figures VI-5 through VI-7)**. Cold climates require an enclosed, insulated shelter.

(a) Snow is the most abundant insulating material.

(b) Air vent is required to prevent carbon monoxide poisoning when using an open flame inside enclosed shelters.

Note: As a general rule, unless you can see your breath, your snow shelter is too warm and should be cooled down to preclude melting and dripping.

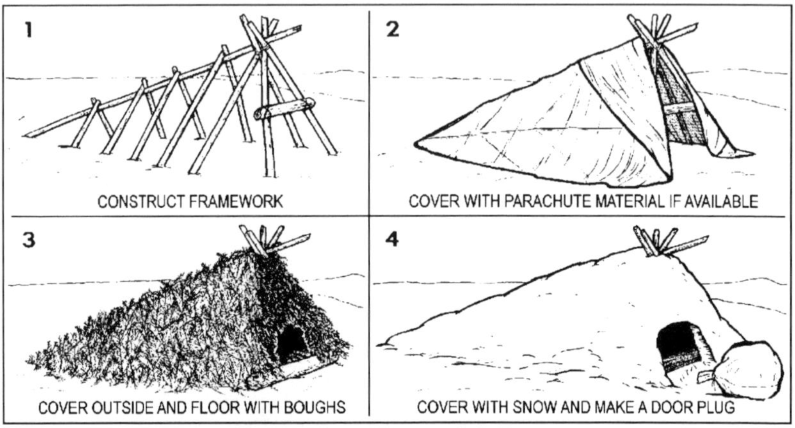

Figure VI-5. Thermal A Frame

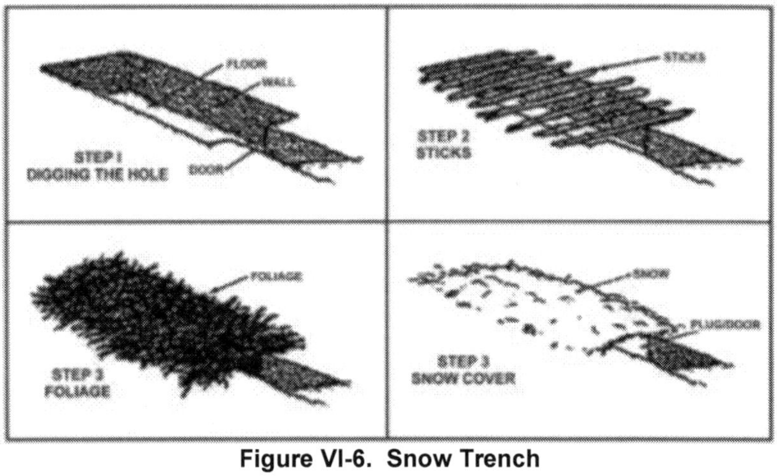

Figure VI-6. Snow Trench

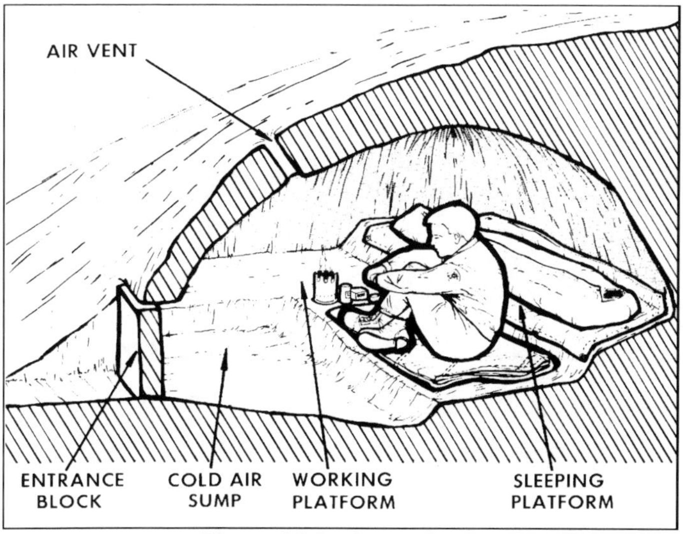

Figure VI-7. Snow Cave

(4) Shade shelter. Hot climates require a shade shelter to protect from ultraviolet rays **(Figure VI-8)**.

(a) To reduce the surface temperature, the shelter floor should be elevated or dug down (approximately 18 inches).

(b) For thermal protection, a minimum of **2** layers of material suspended 12-18 inches above the head is required. White is the best color to reflect heat (inner most layer should be of darker material).

(5) Elevated platform shelter **(Figure VI-9)**. Tropical/wet climates require enclosed, elevated shelter for protection from dampness and insects.

c. Shelter construction.

(1) Have entrance 45-90 degrees from prevailing wind.

(2) Cover with available material.

(a) If natural materials are used, arrange them in layers starting at the bottom with each layer overlapping the previous one. See **Figure VI-10** for an example.

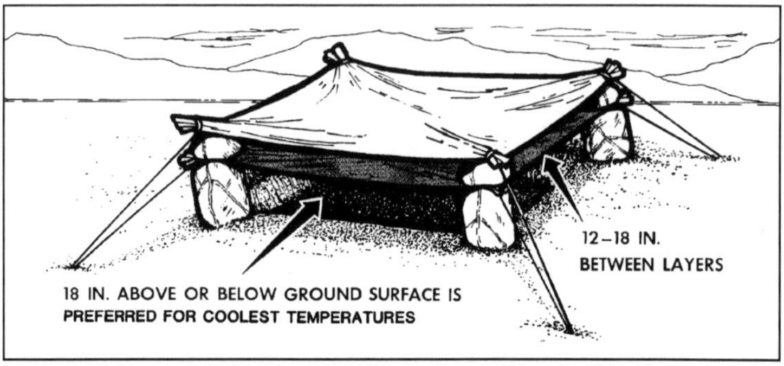

Figure VI-8. Poncho/Parachute Shade Shelter

Figure VI-9. Elevated Platform Shelter

Figure VI-10. Shingle Method

(b) If using porous material like parachute, blankets, etc.—

- Stretch as tight as possible
- Use a 40–60 degree slope.
- Use additional layers in heavy rain.

d. Shelter construction materials:

(1) Raft and raft parts.
(2) Vehicle or aircraft parts.
(3) Blankets, poncho, or parachute material.
(4) Sheet of plastic or plastic bag.
(5) Bark peeled off dead trees.
(6) Boughs, broad leaves, dry moss.
(7) Grass and sod.
(8) Snow.
(9) Sand and rocks.

e. Bed construction. Construct a bed to protect from cold, damp, ground using—

(1) Raft or foam rubber from vehicle seats.
(2) Boughs, leaves, or dry moss.

5. Fires

> **CAUTION:** Weigh hazards and risks of detection against the need for a fire.

a. Evasion considerations:

(1) Use trees or other sources to dissipate smoke.
(2) Use fires at dusk, dawn, or during inclement weather.
(3) Use fires at times when the local populace is cooking.

b. Fire building. The **3** essential elements for starting a fire are heat, fuel, and oxygen.

(1) Heat sources:

(a) Matches or lighter.
(b) Flint and steel (experiment with various rocks and metals until a good spark is produced).
(c) Sparks from batteries.
(d) Concentrated sunlight (use magnifying glass or flashlight reflectors).
(e) Pyrotechnics, such as flares **(last resort)**, etc.

(f) Friction method **(Figure VI-11)**. Without prior training, this method is difficult to master and requires a lot of time to build the device.

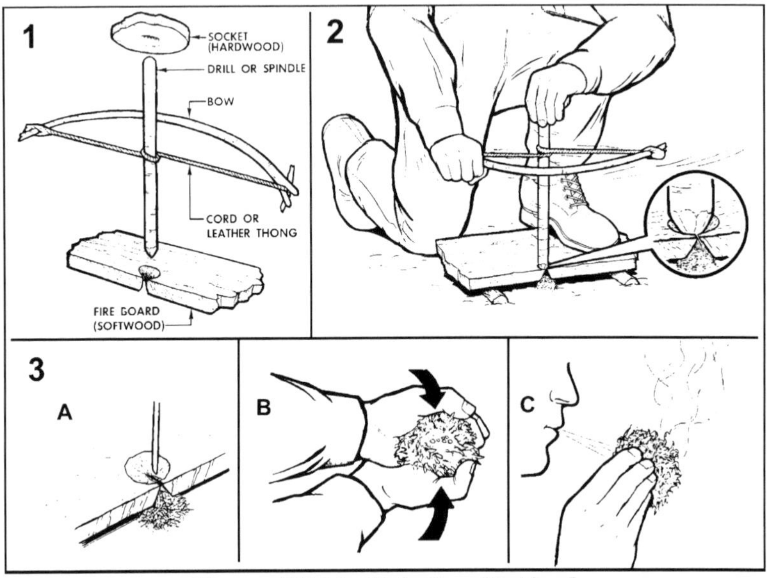

Figure VI-11. Friction Method

> **Note:** If possible, carry a fire-starting device with you.

(2) Fuel is divided into **3** categories: tinder, kindling, and fuel. (Gather large amounts of each category before igniting the fire.)

(a) **Tinder**. Tinder must be very finely shaved or shredded to provide a low combustion point and fluffed to allow oxygen to flow through. (To get tinder to burn hotter and longer, saturate with Vaseline, Chapstick, insect repellant, aircraft fuel, etc.) Examples of tinder include—

- Cotton.
- Candle (shred the wick, not the wax).
- Plastic spoon, fork, or knife.
- Foam rubber.

•Dry bark.
•Dry grasses.
•Gun powder.
•Pitch.
•Petroleum products.

(b) **Kindling**. Kindling must be small enough to ignite from the small flame of the tinder. Gradually add larger kindling until arriving at the size of fuel to burn.

(c) **Fuel**. Examples of fuel include—

•Dry hardwood (removing bark reduces smoke).
•Bamboo (open chambers to prevent explosion).
•Dry dung.

c. Types. Fires are built to meet specific needs or uses.

(1) Tepee fire **(Figure VI-12)**. Use the tepee fire to produce a concentrated heat source for cooking, lighting, or signaling.

Figure VI-12. Tepee Fire

(2) Log cabin fire **(Figure VI-13)**. Use the log cabin fire to produce large amounts of light and heat, to dry out wet wood, and provide coals for cooking, etc.

Figure VI-13. Log Cabin or Pyramid Fires

(3) Sod fire and reflector (**Figure VI-14**). Use fire reflectors to get the most warmth from a fire. Build fires against rocks or logs.

CAUTION: *DO NOT* use porous rocks or riverbed rock—they may explode when heated.

Figure VI-14. Sod Fire and Reflector

(4) Dakota fire hole (Figure VI-15). Use the Dakota fire hole for high winds or evasion situations.

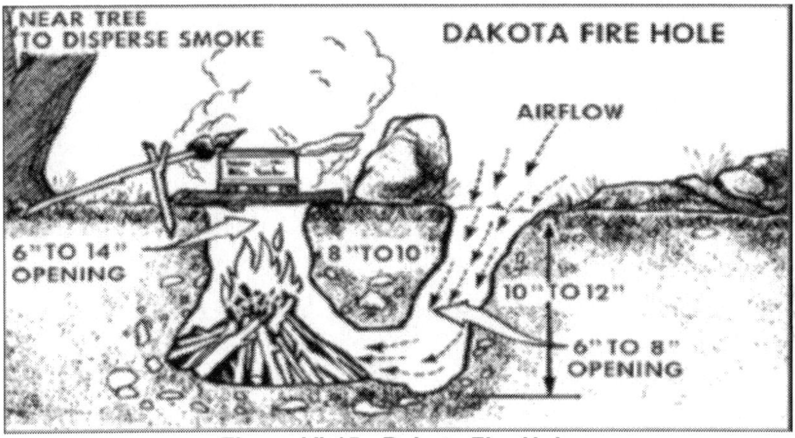

Figure VI-15. Dakota Fire Hole

(5) Improvised stoves **(Figure VI-16)**. These are very efficient.

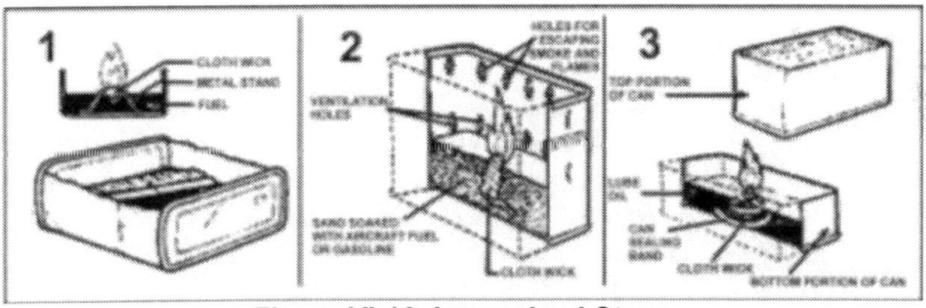

Figure VI-16. Improvised Stove

Chapter VII

Water

1. Water Requirements

Drink **extra** water. Minimum 2 quarts per day to maintain fluid level. Exertion, heat, injury, or an illness increases water loss.

> **Note:** Pale yellow urine indicates adequate hydration.

2. Water Procurement

a. ***DO NOT*** drink—
- (1) Urine.
- (2) Fish juices.
- (3) Blood.
- (4) Sea water.
- (5) Alcohol.
- (6) Melted water from new sea ice.

b. Water sources:
- (1) Surface water (streams, lakes, and springs).
- (2) Precipitation (rain, snow, dew, sleet) **(FigureVII-1)**.
- (3) Subsurface (wells and cisterns).
- (4) Ground water (when no surface water is available) **(Figure VII-2)**.
  - (a) Abundance of lush green vegetation.
  - (b) Drainages and low-lying areas.
  - (c) ***"V"*** intersecting game trails often point to water.
  - (d) Presence of swarming insects indicates water is near.
  - (e) Bird flight in the early morning or late afternoon might indicate the direction to water.
- (5) Snow or ice.
  - (a) ***DO NOT*** eat ice or snow.
    - •Lowers body temperature.
    - •Induces dehydration.
    - •Causes minor cold injury to lips and mouth.

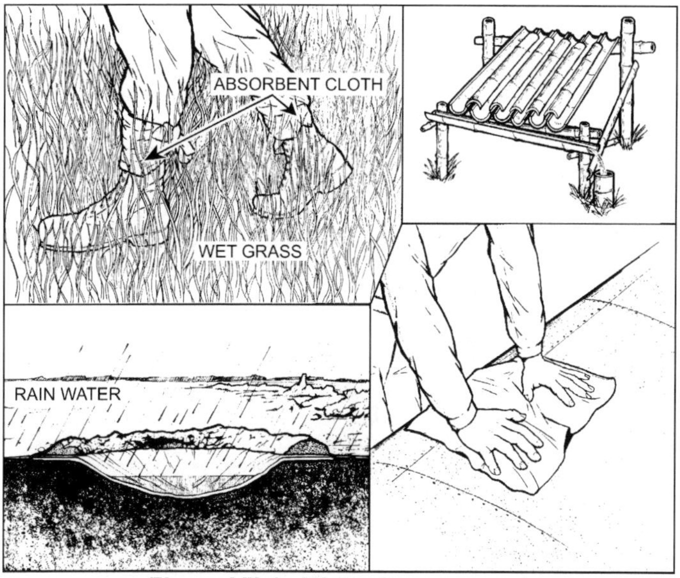

Figure VII-1. Water Procurement

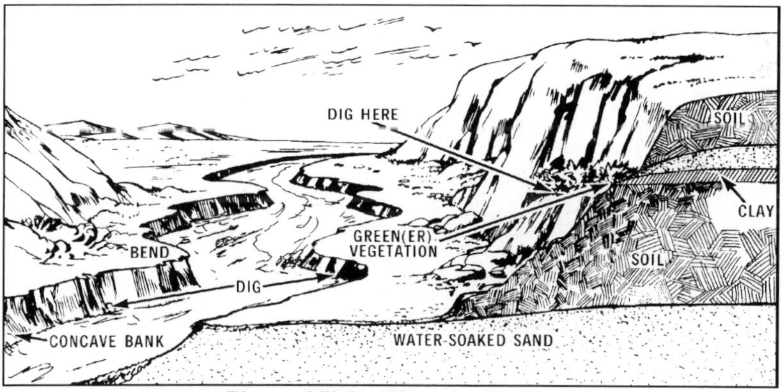

Figure VII-2. Water Indicators

(b) Melt with fire.
- •Stir frequently to prevent damaging container.
- •Speed the process by adding hot rocks or water.

(c) Melt with body heat.
- •Use waterproof container.
- •Place between layers of clothing.
- •*DO NOT* **place next to the skin.**

(d) Use a water generator **(Figure VII-3)**.

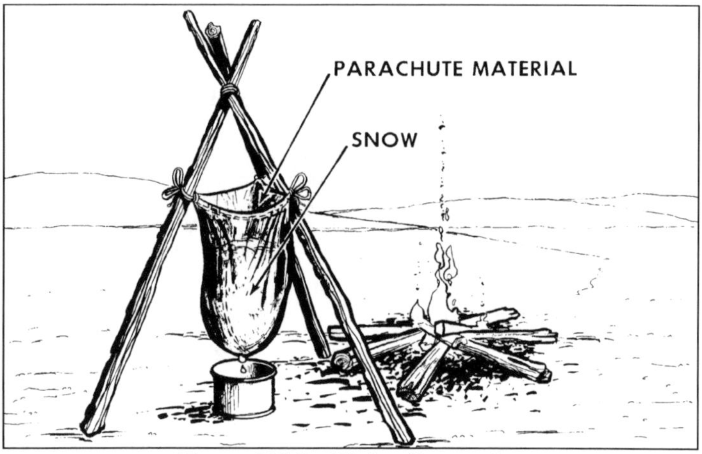

Figure VII-3. Water Generator

(6) Open seas.

(a) Water available in survival kits.

(b) Precipitation.
- •Drink as much as possible.
- •Catch rain in spray shields and life raft covers.
- •Collect dew off raft.

(c) Old sea ice or icebergs **(Table VII-1)**.

Table VII-1. Old Sea Ice or Icebergs

OLD SEA ICE	NEW SEA ICE
Bluish or blackish	Milky or grey
Shatters easily	Does not break easily
Rounded corners	Sharp edges
Tastes relatively salt-free	Tastes extremely salty

(7) Tropical areas.

(a) All open sources previously mentioned.

(b) Vegetation.

•Plants with hollow sections can collect moisture.

•Leaning Tree. Cloth absorbs rain running down tree and drips into container **(Figure VII-4)**.

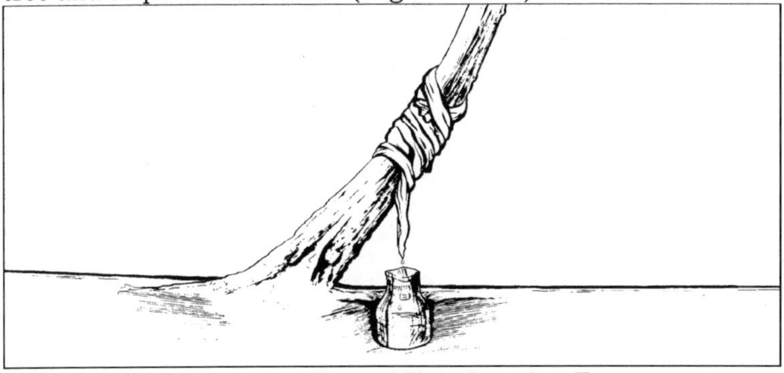

Figure VII-4. Leaning Tree

•Banana plants.

•Water trees (avoid milky sap).

••Tap before dark. Let sap stop running and harden during the daytime.

••Produce most water at night.

••For evasion situations, bore into the roots and collect water.

•Vines **(Figure VII-5A)**.

••Cut bark (***DO NOT*** use milky sap).

••If juice is clear and water like, cut as large a piece of vine as possible (cut the top first).

••Pour into hand to check smell, color, and taste to determine if drinkable.

••***DO NOT*** touch vine to lips.

••When water flow stops, cut off 6 inches of opposite end, water will flow again.

•Old bamboo.

••Shake and listen for water.

••Bore hole at bottom of section to obtain water.

••Cut out entire section to carry with you.
••Filter and purify.
•Green bamboo **(Figure VII-5B)**.

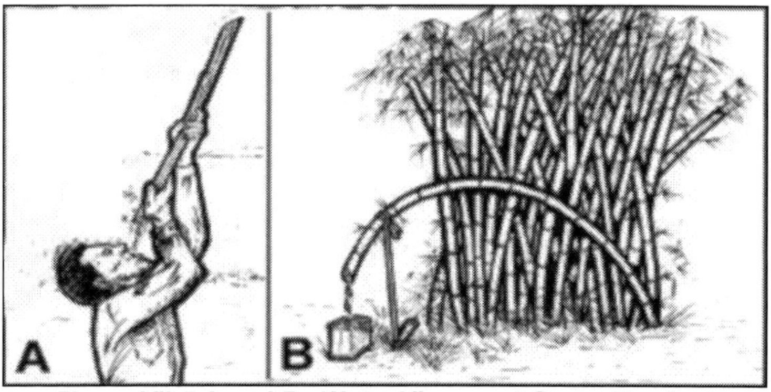

Figure VII-5 A and B. Water Vines and Green Bamboo

CAUTION: Liquid contained in green coconuts (**ripe** coconuts may cause diarrhea).

•Beach well. Along the coast, obtain water by digging a beach well **(Figure VII-6)**.

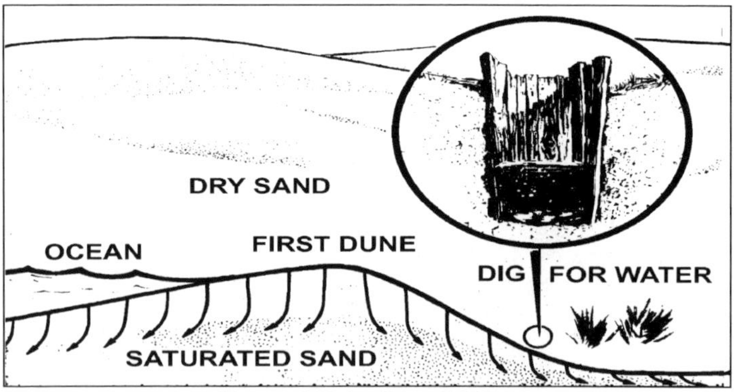

Figure VII-6. Beach Well

(8) Dry areas.

(a) Solar still **(Figure VII-7)**.

(b) Vegetation bag **(Figure VII-8)**.

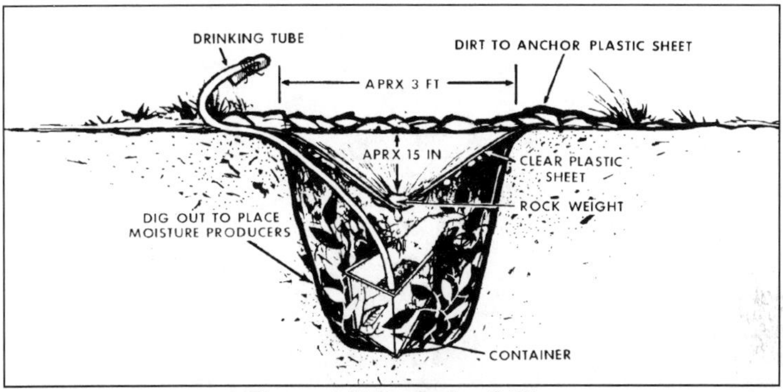

Figure VII-7. Solar Still

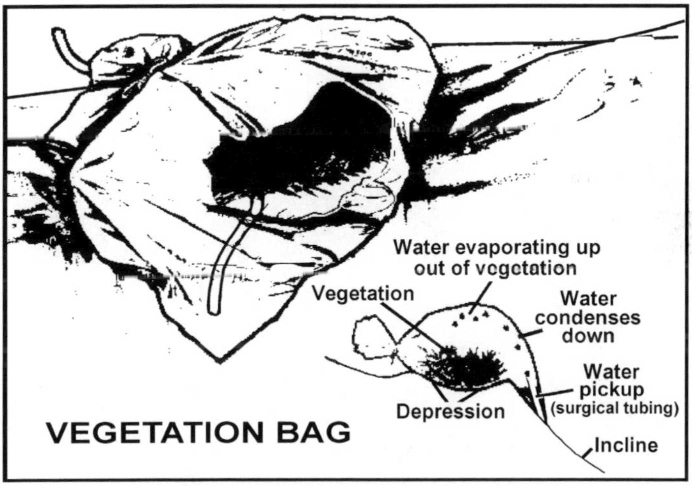

Figure VII-8. Vegetation Bag

(c) Transpiration bag **(Figure VII-9)**.

•Water bag must be clear.

•Water will taste like the plant smells.

(d) Seepage basin **(Figure VII-10)**.

CAUTION: ***DO NOT*** use poisonous/toxic plants in vegetation/transpiration bags.

Figure VII-9. Transpiration Bag

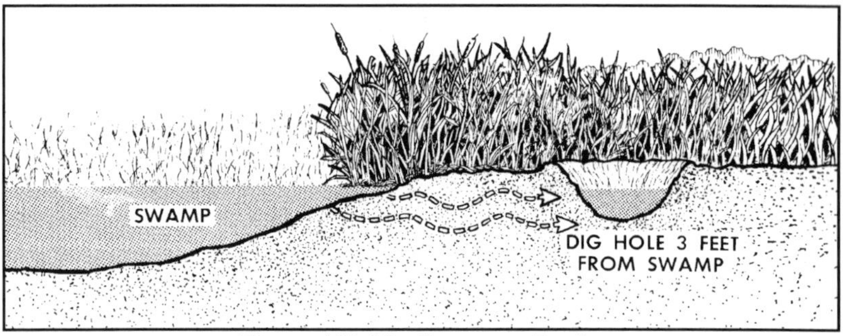

Figure VII-10. Seepage Basin

3. Water Preparation and Storage

a. Filtration. Filter through porous material (sand/charcoal).

b. Purification.

(1) Water from live plants requires no further treatment.

(2) Purify all other water.

(a) Boil at least 1 minute.

(b) Pour from one container to another to improve taste to aerate.

(c) Water purification tablets. Follow instructions on package.

c. Potable Water.

(1) If water cannot be purified, obtain water from a clear, cold, clean, and fast running source (if possible).

(2) Put in clear container and expose to the sun's ultraviolet rays to kill bacteria.

d. Storage. To prevent contamination, use a clean, covered or sealed container.

(1) Trash bag.

(2) Prophylactic.

(3) Section of bamboo.

(4) Flotation gear.

Chapter VIII

FOOD

1. Food Procurement

a. Sources and location.

(1) Mammals can be found where—

(a) Trails lead to watering, feeding, and bedding areas.

(b) Droppings or tracks look fresh.

(2) Birds can be found by—

(a) Observing the direction of flight in the early morning and late afternoon (leads to feeding, watering, and roosting areas).

(b) Listening for bird noises (indication of nesting areas

(3) Fish and other marine life locations **(Figure VIII-1)**.

Figure VIII-1. Fishing Locations

(4) Reptiles and amphibians are found almost worldwide.

(5) Insects are found—

(a) In dead logs and stumps.

(b) At ant and termite mounds.

(c) On ponds, lakes, and slow moving streams.

b. Procurement techniques.

(1) Snares—

(a) Work while unattended.

(b) Location:

•Trails leading to water, feeding, and bedding areas.

•Mouth of dens **(Figure VIII-2)**.

Figure VIII-2. Snare Placement

(c) Construction of simple loop snare.

•Use materials that will not break under the strain of holding an animal.

•Use a figure 8 (locking loop) if wire is used **(Figure VIII-3)**.

••Once tightened, the wire locks in place, preventing reopening, and the animal's escape.

•To construct a squirrel pole **(Figure VIII-4)** use simple loop snares.

•Make noose opening slightly larger than the animal's head **(3-finger** width for squirrels, **fist-sized** for rabbits).

(d) Placement of snares (set as many as possible).

•Avoid disturbing the area.

•Use funneling (natural or improvised) **(Figure VIII-5)**.

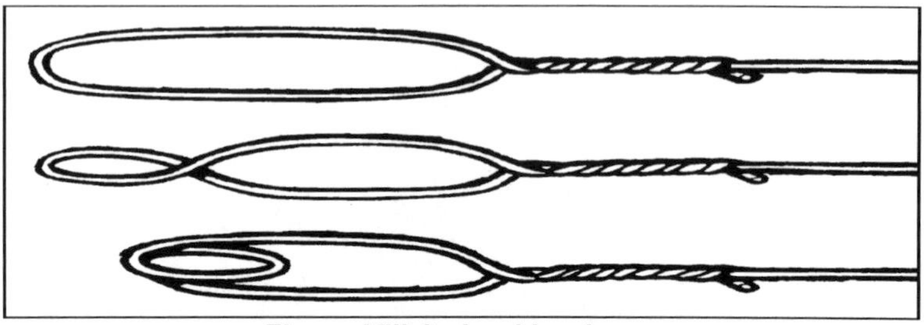

Figure VIII-3. Locking Loop

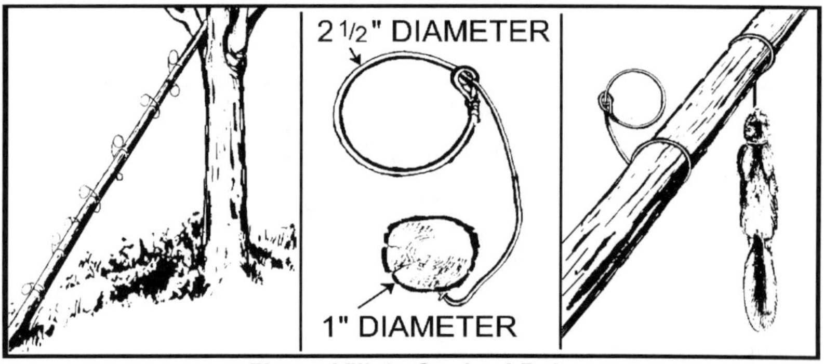

Figure VIII-4. Squirrel Pole

Figure VIII-5. Funneling

(2) Noose stick (easier and safer to use than the hands).

(3) Twist stick **(Figure VIII-6)**.

(a) Insert forked stick into a den until something soft is met.

(b) Twist the stick, binding the animal's hide in the fork.

(c) Remove the animal from the den.

(d) Be ready to **kill** the animal; **it may be dangerous**.

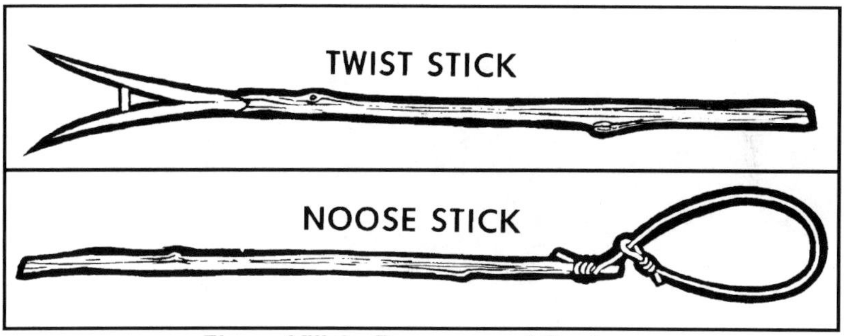

Figure VIII-6. Procurement Devices

(4) Hunting and fishing devices. (See **Figure VIII-7** for fishing procurement methods.)

(a) Club or rock.

(b) Spear.

(c) Slingshot.

(d) Pole, line, and hook.

(e) Net.

(f) Trap.

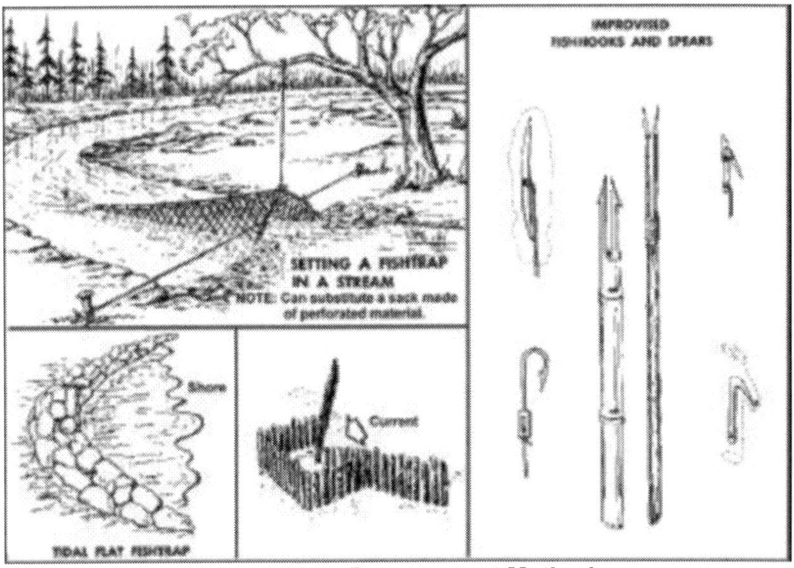

Figure VIII-7. Procurement Methods

(5) Precautions:

(a) Wear shoes to protect the feet when wading in water.

(b) Avoid reaching into dark holes.

(c) **Kill** animals before handling. Animals in distress may attract the enemy.

(d) ***DO NOT*** secure fishing lines to yourself or the raft.

(e) **Kill** fish before bringing them into the raft.

(f) ***DO NOT*** eat fish with—

- Spines.
- Unpleasant odor.
- Pale, slimy gills.
- Sunken eyes.
- Flabby skin.
- Flesh that remains dented when pressed.

(g) ***DO NOT*** eat fish eggs or liver (entrails).

(h) Avoid all crustaceans above the high tide mark.

(i) Avoid cone-shaped shells **(Figure VIII-8)**.

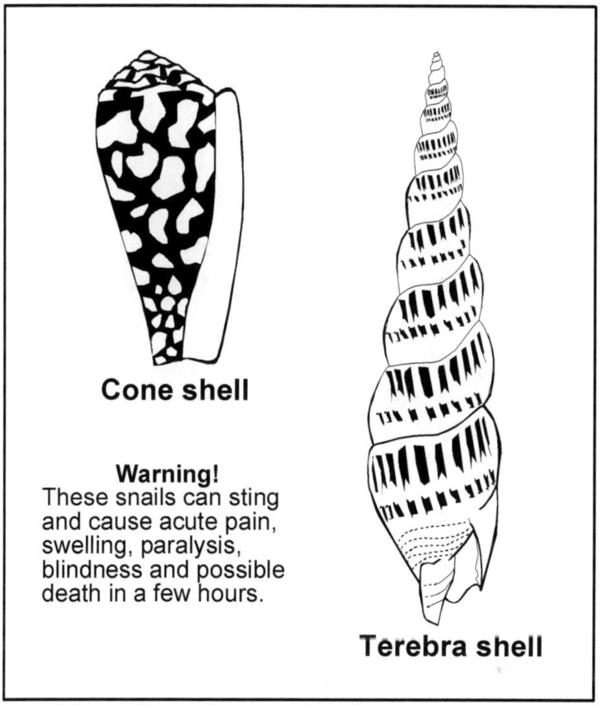

Figure VIII-8. Cone-Shaped Shells of Venomous Snails

(j) Avoid hairy insects; the hairs could cause irritation or infection.

(k) Avoid poisonous insects, for example:
- •Centipedes.
- •Scorpions.
- •Poisonous spiders.

(l) Avoid disease carrying insects, such as—
- •Flies.
- •Mosquitoes.
- •Ticks.

c. Plant Foods. ***Before using the following guide use your evasion chart to identify edible plants:***

Note: If you cannot positively identify an edible plant and choose to try an unknown plant, these guidelines may help determine edibility.

(1) Selection criteria.

(a) Before testing for edibility, ensure there are enough plants to make testing worth your time and effort. Each part of a plant (roots, leaves, stems, bark, etc.) requires more than 24 hours to test. ***DO NOT*** waste time testing a plant that is not abundant.

(b) Test only **1** part of **1** plant at a time.

(c) Remember that eating large portions of plant food on an empty stomach may cause diarrhea, nausea, or cramps. ***Two*** good examples are ***green apples*** and ***wild onions***. Even after testing food and finding it safe, eat in moderation.

(2) Avoid plants with the following characteristics:

Note: Using these guidelines in selecting plants for food may eliminate some edible plants; however, these guidelines will help prevent choosing potentially toxic plants.

(a) Milky sap (dandelion has milky sap but is safe to eat and easily recognizable).

(b) Spines, fine hairs, and thorns (skin irritants/contact dermatitis). ***Prickly pear*** and ***thistles*** are exceptions. ***Bracken fern fiddleheads*** also violate this guideline.

(c) Mushrooms and fungus.

(d) Umbrella shaped flowers (hemlock is eliminated).

(e) Bulbs (***only*** onions smell like onions).

(f) Grain heads with pink, purplish, or black spurs.

(g) Beans, bulbs, or seeds inside pods.

(h) Old or wilted leaves.

(i) Plants with shiny leaves.

(j) White and yellow berries. (Aggregate berries such as black and dewberries are always edible, test all others before eating.)

(k) Almond scent in woody parts and leaves.

d. Test procedures.

> **CAUTION:** Test all parts of the plant for edibility. Some plants have both edible and inedible parts. **NEVER ASSUME** a part that proved edible when cooked is edible raw, test the part raw before eating. The same part or plant may produce varying reactions in different individuals.

(1) Test only **1** part of a plant at a time.

(2) Separate the plant into its basic components (stems, roots, buds, and flowers).

(3) Smell the food for strong acid odors. Remember, smell alone does not indicate a plant is edible or inedible.

(4) ***DO NOT*** eat 8 hours before the test and drink only purified water.

(5) During the 8 hours you abstain from eating, test for contact poisoning by placing a piece of the plant on the inside of your elbow or wrist. The sap or juice should contact the skin. Usually 15 minutes is enough time to allow for a reaction.

(6) During testing, take ***NOTHING*** by mouth **EXCEPT** purified water and the plant you are testing.

(7) Select a small portion of a single part and prepare it the way you plan to eat it.

(8) Before placing the prepared plant in your mouth, touch a small portion (a pinch) to the outer surface of your lip to test for burning or itching.

(9) If after 3 minutes there is no reaction on your lip, place the plant on your tongue and hold it for 15 minutes.

(10) If there is no reaction, thoroughly chew a pinch and hold it in your mouth for 15 minutes **(*DO NOT* SWALLOW)**. If any ill effects occur, rinse out your mouth with water.

(11) If nothing abnormal occurs, swallow the food and wait 8 hours. If **any ill effects** occur during this period, **induce** vomiting and drink a water and charcoal mixture.

(12) If no ill effects occur, eat **¼ cup** of the same plant prepared the same way. Wait another 8 hours. If no ill effects occur, the plant part as prepared is safe for eating.

CAUTION:
1. Ripe tropical fruits should be peeled and eaten raw. Softness, rather than color, is the best indicator of ripeness. Cook unripe fruits and discard seeds and skin.
2. Cook underground portions when possible to reduce bacterial contamination and ease digestion of their generally high starch content.
3. During evasion, you may not be able to cook. Concentrate your efforts on leafy green plants, ripe fruits, and above ground ripe vegetables not requiring significant preparation.

2. Food Preparation

Animal food gives the greatest food value per pound.

a. Butchering and skinning.

(1) Mammals.

(a) Remove the skin and save for other uses.

(a) One cut skinning of small game **(Figure VIII-9)**.

•Open the abdominal cavity.

•Avoid rupturing the intestines.

•Remove the intestines.

•Save inner organs (heart, liver, and kidneys) and all meaty parts of the skull, brain, tongue, and eyes.

(b) Wash when ready to use.

(c) If preserving the meat, remove it from the bones.

(d) Unused or inedible organs and entrails may be used as bait for other game.

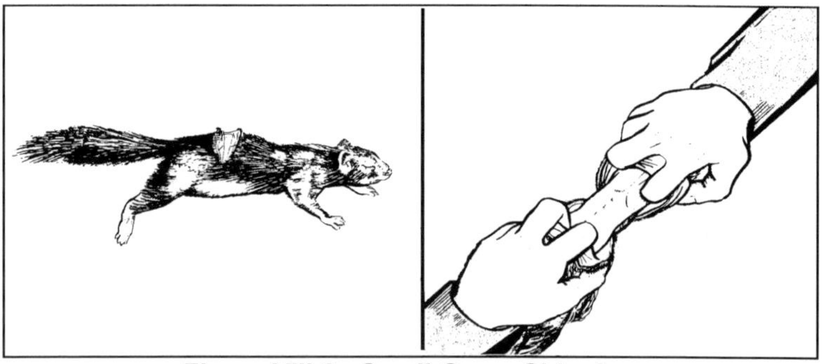

Figure VIII-9. Small Game Skinning

(2) Frogs and snakes.

(a) Skin.

(b) Discard skin, head with 2 inches of body, and internal organs.

(3) Fish.

(a) Scale (if necessary) and gut fish soon after it is caught.

(b) Insert knifepoint into anus of fish and cut open the belly.

(c) Remove entrails.

(d) Remove gills to prevent spoilage.

(4) Birds.

(a) Gut soon after killing.

(b) Protect from flies.

(c) Skin or pluck them.

(d) Skin scavengers and sea birds.

(5) Insects.

(a) Remove all hard portions such as the legs of grasshoppers or crickets. (The rest is edible.)

(b) Recommend cooking grasshopper-size insects.

CAUTION: Dead insects spoil rapidly, ***DO NOT*** save.

(6) Fruits, berries, and most nuts can be eaten raw.

b. Cooking.

CAUTION: To kill parasites, thoroughly cook all wild game, freshwater fish, clams, mussels, snails, crawfish, and scavenger birds. Saltwater fish may be eaten raw.

(1) Boiling (most nutritious method of cooking—drink the broth).

(a) Make metal cooking containers from ration cans.

(b) Drop heated rocks into containers to boil water or cook food.

(2) Baking.

(a) Wrap in leaves or pack in mud.

(b) Bury food in dirt under coals of fire.

(3) Leaching. Some nuts (acorns) must be leached to remove the bitter taste of tannin. Use one of the following leaching methods:

(a) First method:

•Soaking and pouring the water off.

•Crushing and pouring water through. Cold water should be tried first; however, boiling water is sometimes best.

•Discarding water.

(b) Second method:

•Boil, pour off water, and taste the plant.

•If bitter, repeat process until palatable.

(4) Roasting.

(a) Shake shelled nuts in a container with hot coals.

(b) Roast thinly sliced meat and insects over a candle.

3. Food Preservation

b. Keeping an animal alive.

c. Refrigerating.

(1) Long term.

(a) Food buried in snow maintains a temperature of approximately 32 degrees F.

(b) Frozen food will not decompose (freeze in meal-size portions).

(2) Short term.

(a) Food wrapped in waterproof material and placed in a stream remains cool in summer months.

(b) Earth below the surface, particularly in shady areas or along streams, is cooler than the surface.

(c) Wrap food in absorbent material such as cotton and re-wet as the water evaporates.

c. Drying and smoking removes moisture and preserves food.

(1) Use salt to improve flavor and promote drying.

(2) Cut or pound meat into thin strips.

(3) Remove fat.

(4) ***DO NOT*** use pitch woods such as fir or pine; they produce soot giving the meat an undesirable taste.

d. Protecting meat from animals and insects.

(1) Wrapping food.

(a) Use clean material.

(b) Wrap pieces individually.

(c) Ensure all corners of the wrapping are insect proof.

(d) Wrap soft fruits and berries in leaves or moss.

(2) Hanging meat.

(a) Hang meat in the shade.

(b) Cover during daylight hours to protect from insects.

(3) Packing meat on the trail.

(a) Wrap before flies appear in the morning.

(b) Place meat in fabric or clothing for insulation.

(c) Place meat inside the pack for carrying. Soft material acts as insulation helping keep the meat cool.

(d) Carry shellfish, crabs, and shrimp in wet seaweed.

e. ***DO NOT*** store food in the shelter; it attracts unwanted animals.

Chapter IX

INDUCED CONDITIONS

(NUCLEAR, BIOLOGICAL, AND CHEMICAL CONSIDERATIONS)

1. Nuclear Conditions

CAUTION: Radiation protection depends on time of exposure, distance from the source, and shielding.

a. Protection.

(1) **FIND PROTECTIVE SHELTER IMMEDIATELY!**

(2) Gather all equipment for survival (time permitting).

(3) Avoid detection and capture.

(a) Seek existing shelter that may be improved **(Figure IX-1)**.

Figure IX-1. Immediate Action Shelter

(b) If no shelter is available, dig a trench or foxhole as follows:

•Dig trench deep enough for protection, then enlarge for comfort **(Figure IX-2)**.

•Cover with available material.

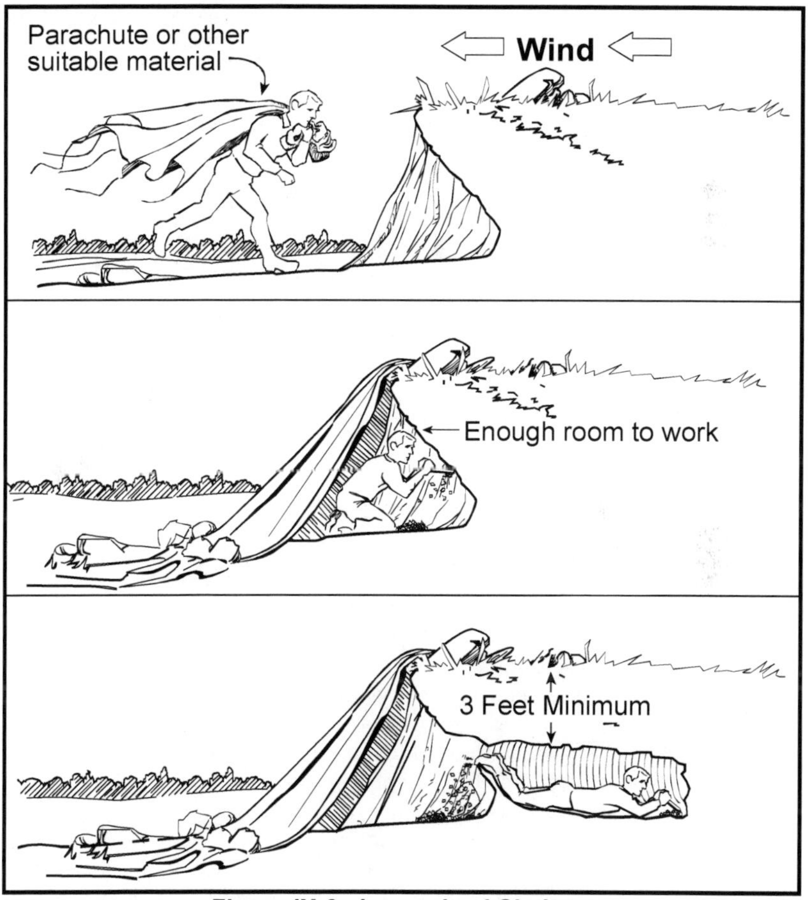

Figure IX-2. Improvised Shelter

(4) Radiation shielding efficiencies **(Figure IX-3).**

NUCLEAR EXPLOSIONS: Fall flat. Cover exposed body parts. Present minimal profile to direction of blast. ***DO NOT** look at fireball!* Remain prone until blast effects are over.

SHELTER: Pick, as soon as possible, 5 minutes unsheltered is maximum!

Priority:
(1) Cave or tunnel covered with 3 or more feet of earth.
(2) Storm/storage cellars
(3) Culverts.
(4) Basements.
(5) Abandoned stone/mud buildings.
(6) Foxhole 4 feet deep (remove topsoil within 2 feet radius of foxhole lip).

RADIATION SHIELDING EFFICIENCIES

Iron/Steel	.7 inches	**Cinder Block**	5.3 inches
Brick	2.0 inches	**Ice**	6.8 inches
Concrete	2.2 inches	**Wood (Soft)**	8.8 inches
Earth	3.3 inches	**Snow**	20.3 inches

One thickness reduces received radiation dose by 1/2. Additional thickness added to any amount of thickness reduces received radiation dose by 1/2.

SHELTER SURVIVAL: Keep contaminated materials out of shelter.
Good Weather: Bury contaminated clothing outside of shelter (recover later).
Bad Weather: Shake strongly or beat with branches. Rinse and /or shake wet clothing. ***DO NOT** wring out!*

PERSONAL HYGIENE: Wash entire body with soap and *any* water; give close attention to fingernails and hairy parts.
No Water: Wipe all exposed skin surfaces with clean cloth or uncontaminated soil. Fallout/dusty conditions--keep entire body covered. Keep handkerchief/cloth over mouth and nose. Improvise goggles. ***DO NOT** smoke!*

DAILY RADIATION TIME TABLE for NO RATE METER

4-6	Complete isolation	**9-12**	2-4 hours exposure per day
3-7	Brief exposure (30 minutes maximum)	**13**	Normal movement.
8	Brief exposure (1 hour maximum)		

Figure IX-3. Radiation Shielding Efficiencies

(5) Leave contaminated equipment and clothing near shelter for retrieval after radioactive decay.

(6) Lie down, keep warm, sleep, and rest.

b. Substance:

(1) Water. Allow no more than 30 minutes exposure on **3d** day for water procurement.

(a) Water sources (in order of preference):

- Springs, wells, or underground sources are **safest**.
- Water in pipes/containers in abandoned buildings.
- Snow (**6** or **more inches below** the surface during the fallout).

•Streams and rivers (filtered before drinking).

•Lakes, ponds, pools, etc.

•Water from below the surface (***DO NOT*** stir up the water).

•Use a seep well.

(b) Water preparation **(Figures IX-4 and IX-5)**.

•Filtering through earth removes 99 percent of radioactivity.

•Purify all water sources.

(2) Food.

(a) Processed foods (canned or packaged) are preferred; wash and wipe containers before use.

(b) Animal foods.

•Avoid animals that appear to be sick or dying.

•Skin carefully to avoid contaminating the meat.

•Before cooking, cut meat away from the bone, leaving at least 1/8 inch of meat on the bone.

•Discard all internal organs.

•Cook all meat until **very well** done.

(c) Avoid.

•Aquatic food sources (use only in extreme emergencies because of high concentration of radiation).

•Shells of all eggs (contents will be safe to eat).

•Milk from animals.

(d) Plant foods (in order of preference).

•Plants whose edible portions grow underground (for example, potatoes, turnips, carrots, etc.). Wash and remove skin.

•Edible portions growing above ground that can be washed and peeled or skinned (bananas, apples, etc.).

•Smooth skinned vegetables, fruits, or above ground plants that are not easily peeled or washed.

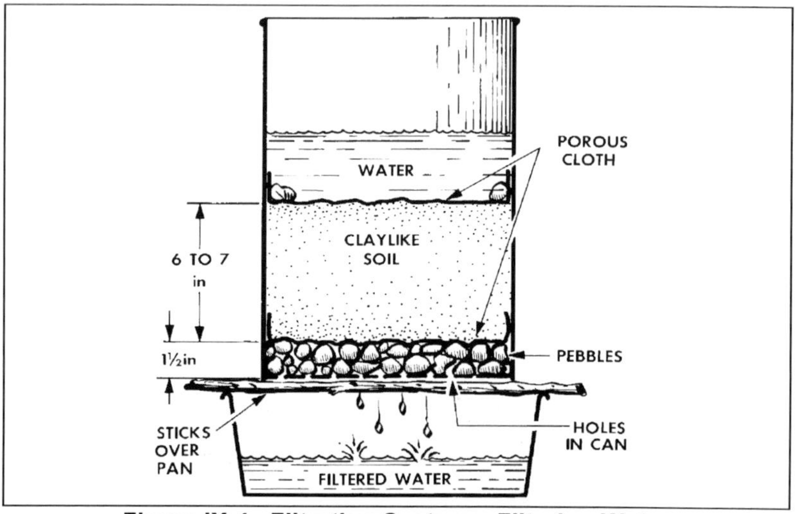

Figure IX-4. Filtration Systems, Filtering Water

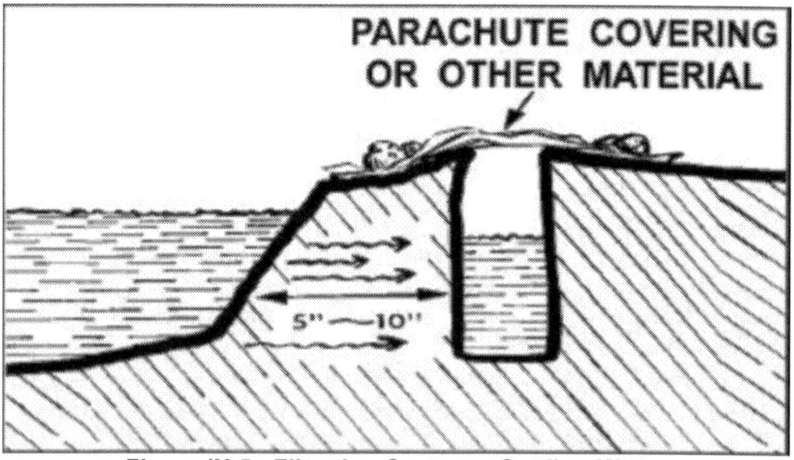

Figure IX-5. Filtration Systems, Settling Water

c. Self-aid:

(1) General rules:

(a) Prevent exposure to contaminants.

(b) Use personal hygiene practices and remove body waste from shelter.

(c) Rest, avoid fatigue.

(d) Drink liquids.

(2) Wounds.

(a) Clean affected area.

(b) Use antibacterial ointment or cleaning solution.

(c) Cover with clean dressing.

(d) Watch for signs of infection.

(3) Burns.

(a) Clean affected area.

(b) Cover with clean dressing.

(4) Radiation sickness (nausea, weakness, fatigue, vomiting, diarrhea, loss of hair, radiation burns).

(a) Time is required to overcome.

(b) Rest.

(c) Drink fluids.

(d) Maintain food intake.

(e) Prevent additional exposure.

2. Biological Conditions

a. Clues which may alert you to a biological attack follow:

(1) Enemy aircraft dropping objects or spraying.

(2) Breakable containers or unusual bombs, particularly those bursting with little or no blast, and muffled explosions.

(3) Smoke or mist of unknown origin.

(4) Unusual substances on the ground or vegetation; sick looking plants or crops.

b. Protection from biological agents follow:

(1) Use protective equipment.

(2) Bathe as soon as the situation permits.

(3) Wash hair and body thoroughly with soap and water.

(4) Clean thoroughly under fingernails.

(5) Clean teeth, gums, tongue, and roof of mouth frequently.

c. Survival tips for biological conditions follow:

(1) Keep your body and living area clean.

(2) Stay alert for clues of biological attack.

(3) Keep nose, mouth, and skin covered.

(4) Keep food and water protected. Bottled or canned foods are safe if sealed. If in doubt, boil food and water for 10 minutes.

(5) Construct shelter in a clear area, away from vegetation, with entrance 90 degrees to the prevailing wind.

(6) If traveling, travel crosswind or upwind (taking advantage of terrain to stay away from depressions).

3. Chemical Conditions

a. Detecting.

(1) Smell. Many agents have little or no odor.

(2) Sight. Many agents are colorless:

(a) Color. Yellow, orange, or red smoke or mist.

(b) Liquid. Oily, dark patches on leaves, ground, etc.

(c) Gas. Some agents appear as a mist immediately after shell burst.

(d) Solid. Most solid state agents have some color.

(3) Sound. Muffled explosions are possible indications of chemical agent bombs.

(4) Feel. Irritation to the nose, eyes, or skin and/or moisture on the skin are danger signs.

(5) Taste. Strange taste in food or water indicates contamination.

(6) General indications. Tears, difficult breathing, choking, itching, coughing, dizziness.

(7) Wildlife. Presence of sick or dying animals.

b. Protection against chemical agents follows:

(1) Use protective equipment.

(2) Avoid contaminated areas.

(a) Exit contaminated area by moving crosswind.

(b) Select routes on high ground.

(c) Avoid cellars, ditches, trenches, gullies, valleys, etc.

(d) Avoid woods, tall grasses, and bushes as they tend to hold chemical agent vapors.

(e) Decontaminate body and equipment as soon as possible by—

•Removing. Pinch-blotting.
•Neutralizing. Warm water.
•Destroying. Burying.

c. Self-aid in chemically contaminated areas.

(1) If a chemical defense ensemble is available—

(a) Use all protective equipment.

(b) Follow antidote directions when needed.

(2) If a chemical defense ensemble is not available—

(a) Remove or tear away contaminated clothing.

(b) Rinse contaminated areas with water.

(c) Improvise a breathing filter using materials available (T-shirt, handkerchief, fabric, etc.).

d. Tips for the survivor:

(1) ***DO NOT*** use wood from a contaminated area for fire.

(2) Look for signs of chemical agents around water sources before procurement (oil spots, foreign odors, dead fish, or animals).

(3) Keep food and water protected.

(4) ***DO NOT*** use plants for food or water in contaminated areas.

Appendix A
THE WILL TO SURVIVE

> **ARTICLE VI CODE OF CONDUCT**
>
> I will never forget that I am an American fighting for freedom, responsible for my actions, and dedicated to the principles which made my country free. I will trust in my God and in the United States of America.

1. Psychology of Survival

a. Preparation—

(1) Know your capabilities and limitations.

(2) Keep a positive attitude.

(3) Develop a realistic plan.

(4) Anticipate fears.

(5) Combat psychological stress by—

(a) Recognizing and anticipating existing *stressors* (injury, death, fatigue, illness, environment, hunger, isolation).

(b) Attributing normal reactions to existing *stressors* (fear, anxiety, guilt, boredom, depression, anger).

(c) Identifying signals of distress created by *stressors* (indecision, withdrawal, forgetfulness, carelessness, and propensity to make mistakes).

b. Strengthen your will to survive with—

(1) The Code of Conduct.

(2) Pledge of Allegiance.

(3) Faith in America.

(4) Patriotic songs.

(5) Thoughts of return to family and friends.

c. Group dynamics of survival include—

(1) Leadership, good organization, and cohesiveness promote high morale:

(a) Preventing panic.

(b) Creating strength and trust in one another.

(c) Favoring persistency in overcoming failure.

(d) Facilitating formulation of group goals.

(2) Taking care of your buddy.

(3) Working as a team.

(4) Reassuring and encouraging each other.

(5) Influencing factors are—
  (a) Enforcing the chain of command.
  (b) Organizing according to individual capabilities.
  (c) Accepting suggestions and criticism.

2. Spiritual Considerations

a. Collect your thoughts and emotions.

b. Identify your personal beliefs.

c. Use self-control.

d. Meditate.

e. Remember past inner sources to help you overcome adversity.

f. Pray for your God's help, strength, wisdom, and rescue.
  (1) Talk to your God.
  (2) Give thanks that God is with you.
  (3) Ask for God's help.
  (4) Pray for protection and a positive outcome.

g. Remember scripture, verses, or hymns; repeat them to yourself and to your God.

h. Worship without aid of written scripture, clergy, or others.

i. Forgive—
  (1) Yourself for what you have done or said that was wrong.
  (2) Those who have failed you.

j. Praise God and give thanks because—
  (1) God is bigger than your circumstances.
  (2) God will see you through (no matter what happens).
  (3) Hope comes from a belief in heaven and/or an after-life.

k. Trust.
  (1) Faith and trust in your God.
  (2) Love for family and self.
  (3) Never lose hope.
  (4) Never give up.

l. With other survivors—
  (1) Identify or appoint a religious lay leader.
  (2) Discuss what is important to you.
  (3) Share scriptures and songs.
  (4) Pray for each other.
  (5) Try to have worship services.

(6) Write down scriptures and songs that you remember.

(7) Encourage each other while waiting for rescue, remember—

(a) Your God loves you.

(b) Praise your God.

Appendix B
Publication Information

1. Scope

This UNCLASSIFIED multiservice tactics, techniques, and procedures publication is designed to assist Service members in a survival situation regardless of geographic location.

2. Purpose

This publication provides Service members a quick reference, weatherproof, pocket-sized guide on basic survival, evasion, and recovery information.

3. Application

The target audience for this publication is any Service member requiring basic survival, evasion, and recovery information.

4. Implementation Plan

Participating Service command offices of primary responsibility (OPRs) will review this publication, validate the information, and reference and incorporate it in Service and command manuals, regulations, and curricula as follows:

Army. The Army will incorporate the procedures in this publication in US Army training and doctrinal publications as directed by the commander, US Army Training and Doctrine Command (TRADOC). Distribution is in accordance with the Department of Army Form 12-99-R.

Marine Corps. The Marine Corps will incorporate the procedures in this publication in US Marine Corps training and doctrinal publications as directed by the commanding general, US Marine Corps Combat Development Command (MCCDC). Distribution is in accordance with Marine Corps Publication Distribution System.

Navy. The Navy will incorporate these procedures in US Navy doctrinal and training publications as directed by the commander, Navy Warfare Development Command (NWDC). Distribution is in accordance with Military Standard Requisitioning and Issue Procedures Desk Guide and Navy Standing Operating Procedures Publication 409.

Air Force. Air Force units will validate and incorporate appropriate procedures in accordance with applicable governing directives. Distribution is in accordance with Air Force Instructions 33-360.

5. User Information

a. The TRADOC-MCCDC-NWDC-AFDC Air Land Sea Application (ALSA) Center developed this publication with the joint participation of the approving Service commands. ALSA will review and update this publication as necessary.

b. This publication reflects current joint and Service doctrine, command and control (C2) organizations, facilities, personnel, responsibilities, and procedures. Changes in Service protocol, appropriately reflected in joint and Service publications, will likewise be incorporated in revisions to this document.

c. We encourage recommended changes for improving this publication. Key your comments to the specific page and paragraph and provide a rationale for each recommendation. Send comments and recommendation directly to—

Army

Commander
US Army Training and Doctrine Command
ATTN: ATDO-A
Fort Monroe VA 23651-5000
DSN 680-3153 COMM (757) 727-3153

Marine Corps

Commanding General
US Marine Corps Combat Development Command
ATTN: C42
3300 Russell Road
Quantico VA 22134-5001
DSN 278-6234 COMM (703) 784-6234

Navy

Commander
Navy Warfare Development Command (Det Norfolk)
ATTN: ALSA Liaison Officer
1530 Gilbert Street
Norfolk VA 23511-2785
DSN 262-2782 COMM (757) 322-2782
E-mail: ndcjoint@nctamslant.navy.mil

Air Force

HQ Air Force Doctrine Center
ATTN: DJ
216 Sweeney Boulevard Suite 109
Langley AFB VA 23665-2722
DSN 754-8091 COMM (757) 764-8091
E-mail Address: afdc.dj@langley.af.mil

ALSA

ALSA Center
ATTN: Director
114 Andrews Street
Langley AFB, VA 23665-2785
DSN 575-0902 COMM (757) 225-0902
E-mail: alsadirector@langley.af.mil

This publication has been prepared under our direction for use by our respective commands and other commands as appropriate.

JOHN N. ABRAMS
General, USA
Commander
Training and Doctrine Command

J. E. RHODES
Lieutenant General, USMC
Commanding General
Marine Corps Combat
Development Command

B. J. SMITH
Rear Admiral, USN
Commander
Navy Warfare Development
Command

TIMOTHY A. KINNAN
Major General, USAF
Commander
Headquarters Air Force
Doctrine Center

This publication is available on the Army Doctrinal and Training Digital Library (ADTDL) at http://155.217.58.58.

FM 21-76-1
MCRP 3-02H
NWP 3-50.3
*AFTTP(I) 3-2.26
29 JUNE 1999

By Order of the Secretary of the Army:

ERIC K. SHINSEKI
General, United States Army
Chief of Staff

Official:

JOEL B. HUDSON
Administrative Assistant to the
Secretary of the Army

By Order of the Secretary of the Air Force:

TIMOTHY A. KINNAN
Major General, USAF
Commander
Headquarters Air Force Doctrine Center

*Supersedes: AFPAM 36-2246, 1 March 1996
Air Force Distribution: F

DISTRIBUTION:
Active Army, Army National Guard, and U.S. Army Reserve: To be distributed in accordance with the initial distribution number 110906, requirements for FM 21-76-1.

MARINE CORPS: PCN 14400006800

Made in the USA
Las Vegas, NV
07 November 2024

11211531R10059